# δ Delta

## Focus: Division

**Test Booklet**

1·888·854·MATH(6284) - MathUSee.com
Sales@MathUSee.com

# Math·U·See®

Sales@MathUSee.com - 1·888·854·MATH(6284) - MathUSee.com



Graphic Design by Christine Minnich



Printed in the United States of America

Fill in the parentheses with the factors, and write the product in the oval. Then write the problem under the rectangle.

1.

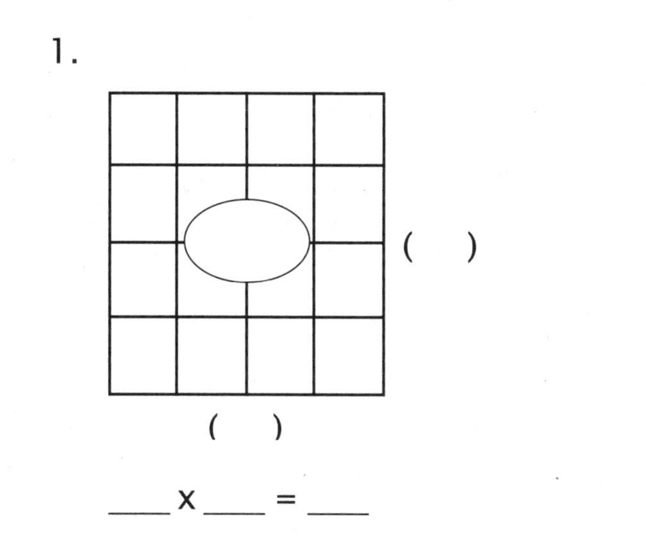

___ x ___ = ___

2.

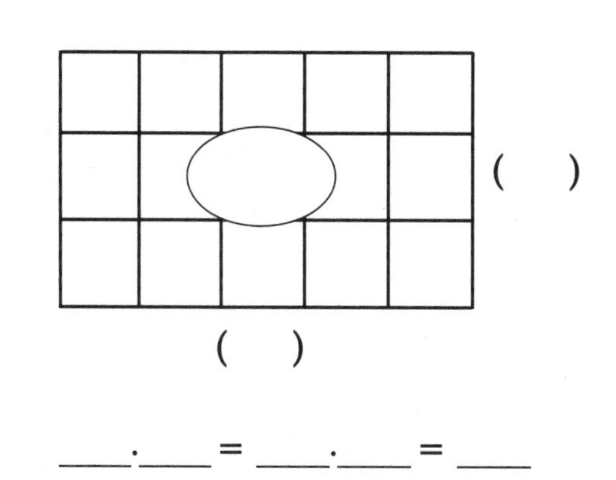

___·___ = ___·___ = ___

Solve for the unknown.

3. $3X = 27$

4. $8Y = 64$

5. $4Q = 20$

6. $6B = 6$

7. $5D = 45$

8. $6F = 24$

9. $10R = 100$

10. $2H = 14$

Find the area.

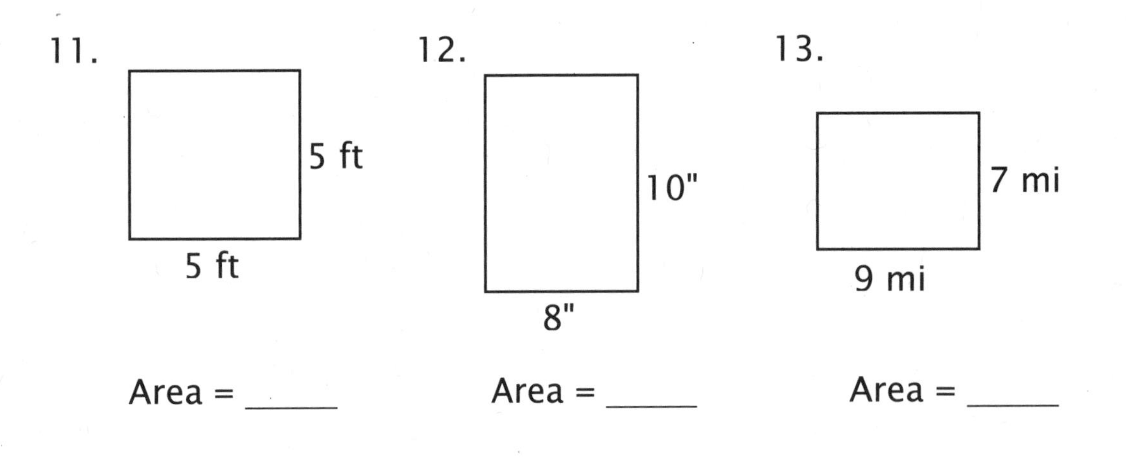


14. Dan earned $90. If he earns $9 an hour, how long had he worked? _____

15. A room measures eight feet by nine feet. What is the area of the room? _____

Divide.

1. $8 \div 1 =$ _____

2. $18 \div 2 =$ _____

3. $6 \div 2 =$ _____

4. $4 \div 1 =$ _____

5. $10 \div 2 =$ _____

6. $8 \div 2 =$ _____

7. $\frac{2}{2} =$ _____

8. $\frac{7}{1} =$ _____

9. $\frac{4}{2} =$ _____

Solve for the unknown.

10. $5X = 15$

11. $5R = 25$

12. $2B = 20$

13. $10Y = 60$

Multiply.

14. $\begin{array}{r} 8 \\ \times\ 5 \\ \hline \end{array}$

15. $\begin{array}{r} 10 \\ \times\ 7 \\ \hline \end{array}$

16. $4 \cdot 5 =$ _____

17. $(10)(3) =$ _____

Find the area. Don't forget to label your answer correctly.

18.

19. Madison divided 12 gumdrops between herself and a friend. How many gumdrops did her friend receive? _____

20. A pie is cut into eight pieces. If there are eight people around the dinner table, how many pieces of pie can each one have? _____

Divide.

1. $10\overline{)20}$

2. $10\overline{)80}$

3. $10\overline{)60}$

4. $10\overline{)90}$

5. $1\overline{)8}$

6. $2\overline{)14}$

7. $2\overline{)16}$

8. $1\overline{)5}$

9. $6 \div 2 =$ ______

10. $10 \div 2 =$ ______

11. $\frac{10}{10} =$ ______

12. $\frac{4}{2} =$ ______

Multiply.

13. $\begin{array}{r} 5 \\ \times\ 3 \\ \hline \end{array}$

14. $\begin{array}{r} 3 \\ \times\ 3 \\ \hline \end{array}$

15. $6 \cdot 3 =$ _____

16. $3 \times 4 =$ _____

17. Ethan bought 10 quarts of oil for his car. How many pints did he buy? _____

18. Bailey found the path to be eight yards long. How many feet long is the path? _____

19. Caleb has $40 to spend on gifts. If he uses $10 for each gift, how many gifts can he buy? _____

20. Meredith drew a rectangle that measured seven inches by nine inches. What was the area of the rectangle? _____

Divide.

1. $5\overline{)15}$

2. $3\overline{)18}$

3. $3\overline{)12}$

4. $5\overline{)45}$

5. $5\overline{)30}$

6. $5\overline{)25}$

7. $3\overline{)27}$

8. $3\overline{)6}$

9. $14 \div 2 =$ _____

10. $50 \div 10 =$ _____

11. $\frac{24}{3} =$ _____

12. $\frac{9}{3} =$ _____

Solve for the unknown.

13. $3X = 18$ 14. $7R = 21$ 15. $5Y = 20$

Add. Regroup when necessary.

16.
```
  2 5
+ 3 8
-----
```

17.
```
  4 7
+ 7 3
-----
```

18.
```
  6 4
+ 1 9
-----
```

19. What is the area of a rectangle that measures five miles by seven miles? _____

20. Janna bought 75 red beads and 69 blue beads for a craft project. How many beads did she buy? _____

Solve for the unknown.

1. $9A = 81$

2. $9X = 18$

3. $9Q = 45$

4. $9T = 72$

5. $9X = 54$

6. $9R = 27$

7. $9B = 63$

8. $9Y = 36$

Divide.

9. $35 \div 5 =$ _____

10. $24 \div 3 =$ _____

11. $\frac{5}{5} =$ _____

12. $\frac{18}{3} =$ _____

13. $3\overline{)21}$

14. $5\overline{)25}$

15. $3\overline{)15}$

16. $2\overline{)16}$

17. Write the symbol for parallel. _____

Draw a pair of lines that look parallel.

18. Write the symbol for perpendicular. _____

Draw a pair of lines that look perpendicular.

19. A tree is 30 feet tall. How many yards tall is the tree? _____

20. Cassie spent $15 for a book and $26 for a gift. How much money did she spend in all? _____

Divide.

1. $9\overline{)18}$

2. $9\overline{)36}$

3. $9\overline{)45}$

4. $9\overline{)81}$

5. $9\overline{)27}$

6. $9\overline{)54}$

7. $9\overline{)72}$

8. $9\overline{)63}$

9. $90 \div 10 =$ ______

10. $10 \div 2 =$ ______

11. $\frac{15}{3} =$ ______

12. $\frac{12}{3} =$ ______

13. $24 \div 3 =$ _____

14. $20 \div 2 =$ _____

15. $\frac{9}{9} =$ _____

16. $\frac{21}{3} =$ _____

Subtract.

17. $\begin{array}{r} 91 \\ -\ 76 \\ \hline \end{array}$

18. $\begin{array}{r} 42 \\ -\ 13 \\ \hline \end{array}$

19. $\begin{array}{r} 80 \\ -\ 35 \\ \hline \end{array}$

20. Hannah made 18 treats yesterday and 17 more today. She plans to divide the treats among her five friends as gifts. How many treats will each friend receive? _____

## UNIT TEST **Lessons 1-6**

Divide.

1. $2\overline{)4}$

2. $3\overline{)24}$

3. $9\overline{)45}$

4. $1\overline{)2}$

5. $3\overline{)18}$

6. $5\overline{)30}$

7. $9\overline{)81}$

8. $10\overline{)60}$

9. $16 \div 2 =$ _____

10. $21 \div 3 =$ _____

11. $\frac{63}{9} =$ _____

12. $\frac{5}{5} =$ _____

13. $27 \div 9 =$ _____

14. $10 \div 2 =$ _____

15. $\frac{40}{10} =$ _____

16. $\frac{12}{3} =$ _____

Divide.

17. $9\overline{)45}$

18. $2\overline{)14}$

19. $5\overline{)25}$

20. $2\overline{)18}$

21. $5\overline{)50}$

22. $3\overline{)9}$

23. $9\overline{)9}$

24. $3\overline{)27}$

25. $90 \div 10 =$ ______

26. $15 \div 3 =$ ______

27. $\frac{12}{2} =$ ______

28. $\frac{4}{1} =$ ______

29. $40 \div 5 =$ ______

30. $54 \div 9 =$ ______

31. $\frac{70}{10} =$ ______

32. $\frac{15}{5} =$ ______

Divide.

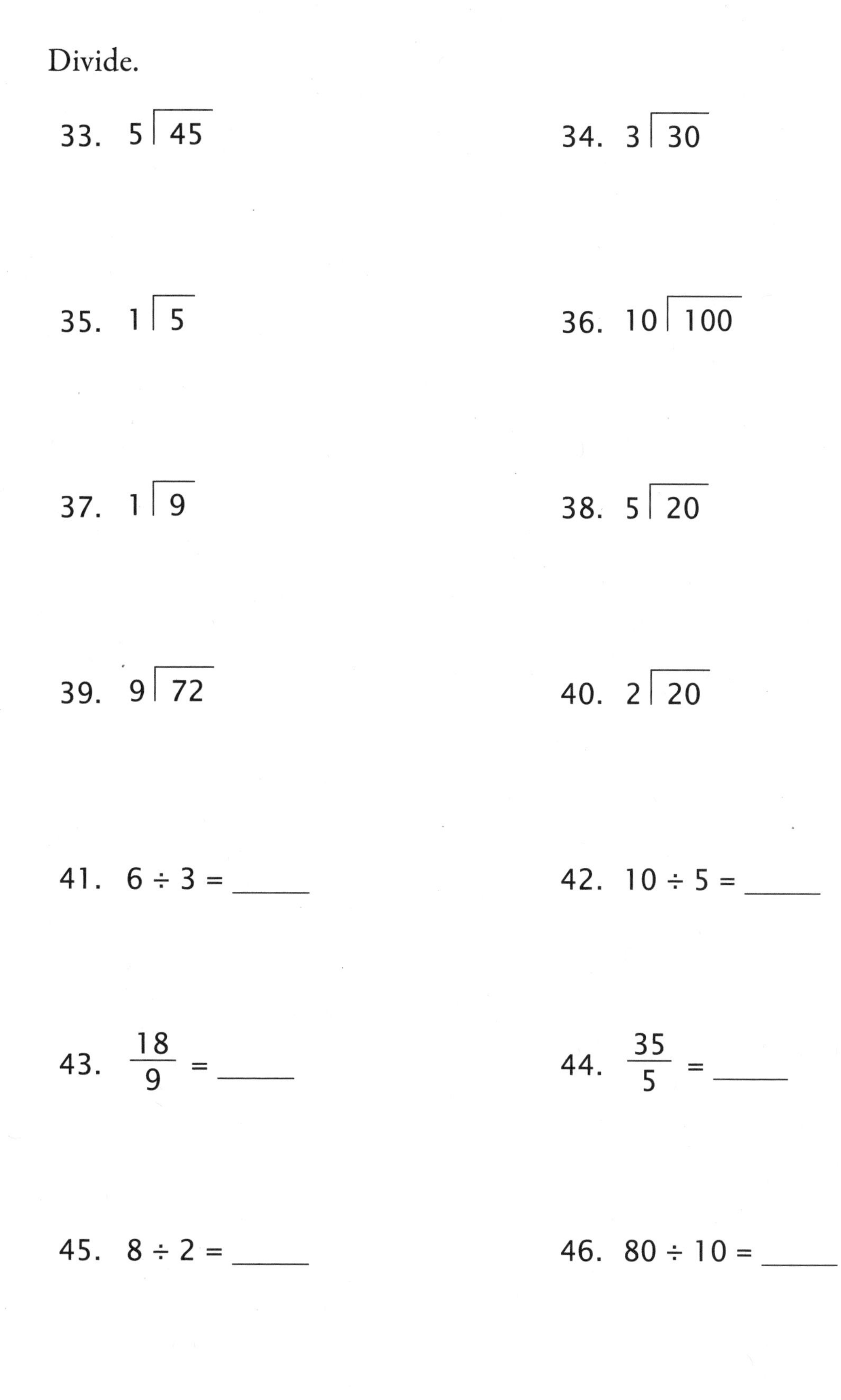

33. $5\overline{)45}$

34. $3\overline{)30}$

35. $1\overline{)5}$

36. $10\overline{)100}$

37. $1\overline{)9}$

38. $5\overline{)20}$

39. $9\overline{)72}$

40. $2\overline{)20}$

41. $6 \div 3 =$ ______

42. $10 \div 5 =$ ______

43. $\frac{18}{9} =$ ______

44. $\frac{35}{5} =$ ______

45. $8 \div 2 =$ ______

46. $80 \div 10 =$ ______

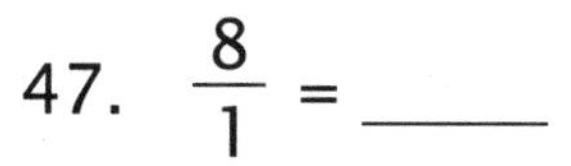

47. $\frac{8}{1} =$ ______

48. $\frac{3}{3} =$ ______

Add or subtract.

49. $\begin{array}{r} 56 \\ +39 \\ \hline \end{array}$

50. $\begin{array}{r} 62 \\ -25 \\ \hline \end{array}$

51. $\begin{array}{r} 81 \\ -46 \\ \hline \end{array}$

52. Draw a pair of lines that look parallel.

53. Draw a pair of lines that look perpendicular.

54. What is the area of a rectangle that measures eight feet by nine feet? _____

55. How many feet are there in five yards? _____

56. How many yards are there in 27 feet? _____

57. How many pints are there in six quarts? _____

58. How many quarts are there in 24 pints? _____

Find the area.

1.

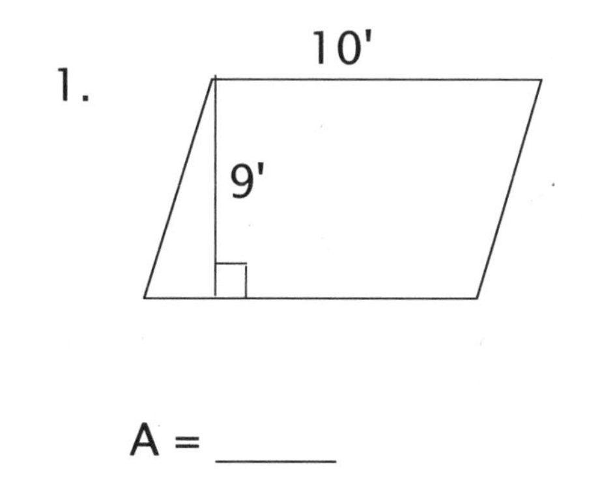


A = ______

2.

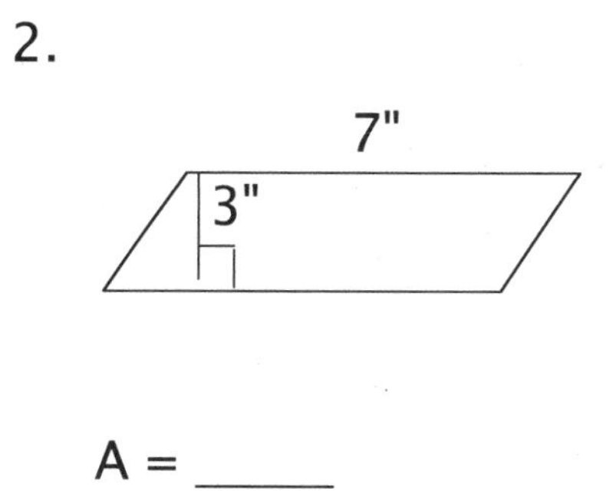


A = ______

Divide.

3. $9\overline{)36}$

4. $3\overline{)18}$

5. $5\overline{)20}$

6. $2\overline{)12}$

7. $81 \div 9 =$ ______

8. $40 \div 10 =$ ______

9. $\frac{4}{2} =$ ______

10. $\frac{63}{9} =$ ______

Solve for the unknown.

11. $6T = 12$

12. $6Y = 30$

13. $6A = 54$

14. $6F = 18$

15. $6X = 42$

16. $6R = 48$

17. $6B = 24$

18. $6Y = 36$

19. Abby raises white rabbits. She had 25 rabbits, but she sold 16 of them. She wants to put her remaining rabbits in cages with three rabbits in each cage. How many cages does she need? _____

20. Rachel bought 18 pint jars. How many quarts of her special homemade jam should she make to fill the jars? _____

Divide.

1. $6\overline{)12}$

2. $6\overline{)24}$

3. $6\overline{)54}$

4. $6\overline{)30}$

5. $6\overline{)42}$

6. $6\overline{)48}$

7. $6\overline{)18}$

8. $6\overline{)36}$

9. $72 \div 9 =$ _____

10. $20 \div 5 =$ _____

11. $\frac{8}{2} =$ _____

12. $\frac{27}{3} =$ _____

Add or subtract.

13. $\begin{array}{r} 23 \\ -\ \ 5 \\ \hline \end{array}$

14. $\begin{array}{r} 72 \\ +\ 19 \\ \hline \end{array}$

15. $\begin{array}{r} 53 \\ -\ 45 \\ \hline \end{array}$

Multiply.

16. $\begin{array}{r} 22 \\ \times\ 13 \\ \hline \end{array}$

17. $\begin{array}{r} 45 \\ \times\ 24 \\ \hline \end{array}$

18. $\begin{array}{r} 16 \\ \times\ 37 \\ \hline \end{array}$

19. Jeremy was bored, so he counted people's feet as they walked by. If he counted 20 feet, how many people had gone by? ______

20. A parallelogram has an area of 36 square feet. If the height is six feet, what is the length of the base? _____

Find the area.

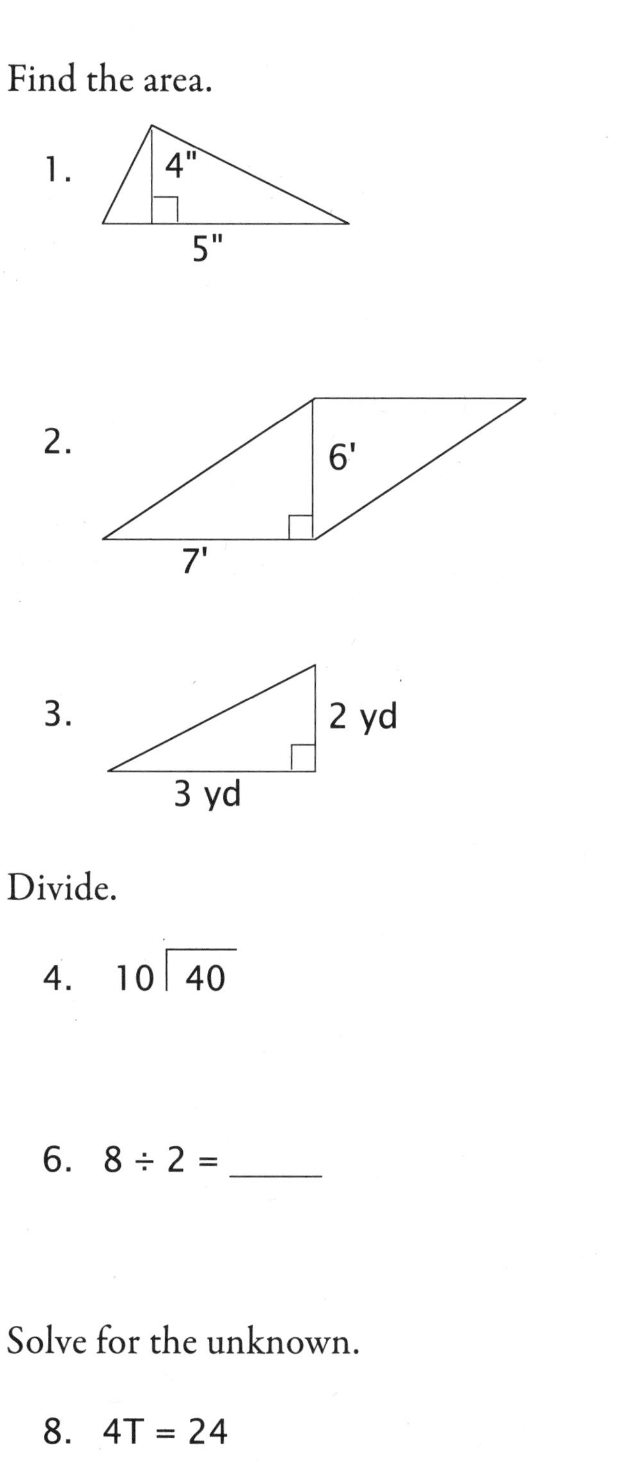


1. A = ____

2. A = ____

3. A = ____

Divide.

4. $10\overline{)40}$

5. $3\overline{)12}$

6. $8 \div 2 =$ ____

7. $\frac{45}{5} =$ ____

Solve for the unknown.

8. 4T = 24

9. 4Y = 32

10. 4A = 16

11. 4F = 28

Multiply.

12. $\begin{array}{r} 84 \\ \times 22 \\ \hline \end{array}$

13. $\begin{array}{r} 43 \\ \times 35 \\ \hline \end{array}$

14. $\begin{array}{r} 67 \\ \times 54 \\ \hline \end{array}$

Add. Make 10 when possible.

15. $\begin{array}{r} 25 \\ 15 \\ 24 \\ +61 \\ \hline \end{array}$

16. $\begin{array}{r} 44 \\ 38 \\ 62 \\ 56 \\ +11 \\ \hline \end{array}$

17. $\begin{array}{r} 90 \\ 23 \\ 57 \\ 18 \\ +82 \\ \hline \end{array}$

18. Can a triangle have two parallel sides?_____

19. Timothy counted animals at the fair. He saw 25 cows, 16 horses, 18 pigs, and 32 sheep. How many animals did he see in all?_____

20. Each of the animals Timothy saw (#19) had four hooves. How many hooves did he see altogether?_____

Divide.

1. $4\overline{)8}$

2. $4\overline{)32}$

3. $4\overline{)16}$

4. $4\overline{)12}$

5. $28 \div 4 =$ _____

6. $20 \div 4 =$ _____

7. $\frac{36}{4} =$ _____

8. $\frac{24}{4} =$ _____

9. $18 \div 6 =$ _____

10. $42 \div 6 =$ _____

11. $\frac{30}{6} =$ _____

12. $\frac{48}{6} =$ _____

Follow the signs.

13. $\begin{array}{r} 71 \\ 34 \\ 59 \\ +\ 26 \\ \hline \end{array}$

14. $\begin{array}{r} 65 \\ -\ 39 \\ \hline \end{array}$

15. $\begin{array}{r} 84 \\ \times\ 62 \\ \hline \end{array}$

Find the area.

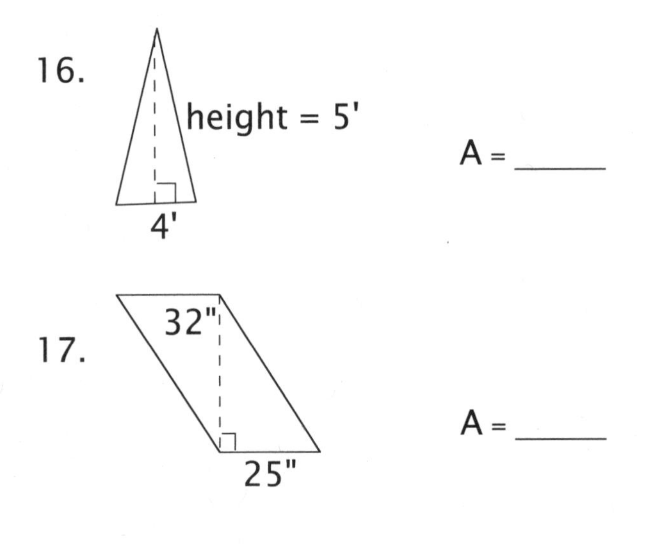


16. A = ______

17. A = ______

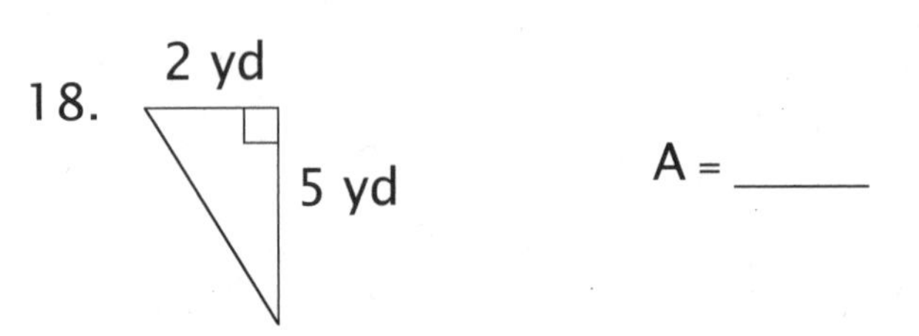


18. A = ______

19. Bria brought 24 quarts of lemonade to the picnic. How many gallons was that? _____

20. Alaina has 40 quarters in her piggy bank. How many dollars does she have? _____

Find the average of the given numbers.

1. 2, 7, 8, 3 Average = _____

2. 4, 3, 6, 7, 10 Average = _____

3. 5, 11, 14 Average = _____

Solve for the unknown.

4. $7A = 49$

5. $7X = 42$

6. $7B = 56$

7. $8D = 64$

8. $8Q = 56$

9. $8G = 72$

Divide.

10. $9\overline{)63}$

11. $6\overline{)48}$

12. $5\overline{)40}$

13. $21 \div 3 =$ _____

14. $32 \div 4 =$ _____

15. $\frac{80}{10} =$ _____

Find the area.

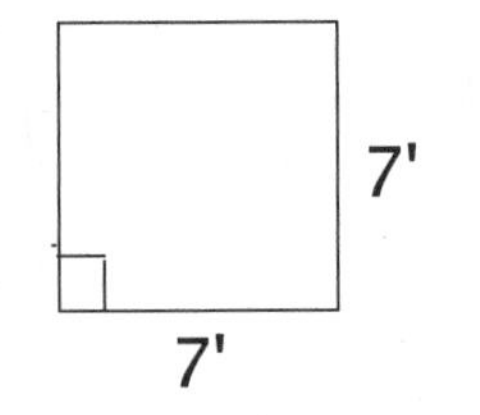


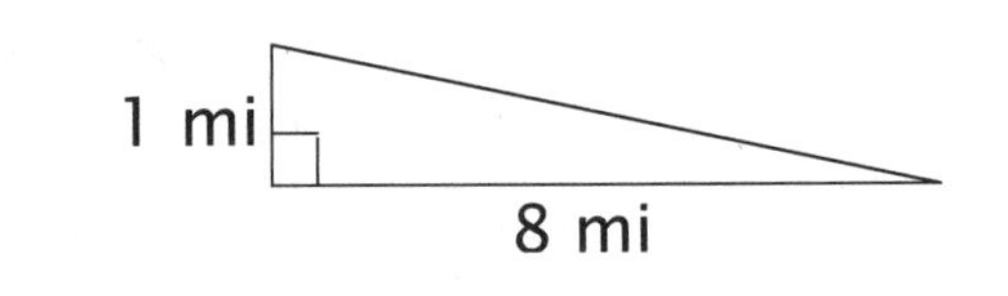


16. A = _____

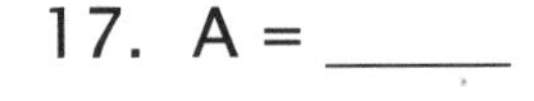

17. A = _____

18. Can a triangle have a pair of sides that are perpendicular to each other? _____

19. Sarah bought 11 feet of red ribbon and 16 feet of yellow ribbon. How many *yards* of ribbon does she have altogether? _____

20. Alexis is trying to drink more water. She drank three pints on Monday, two pints on Tuesday, four pints on Wednesday, and three pints on Thursday. What is her average for the four days?

_____

How many quarts of water did she drink in all? _____

Divide.

1. $7\overline{)28}$

2. $7\overline{)63}$

3. $8\overline{)56}$

4. $8\overline{)16}$

5. $7\overline{)14}$

6. $7\overline{)35}$

7. $8\overline{)24}$

8. $8\overline{)72}$

9. $48 \div 8 =$ ______

10. $42 \div 7 =$ ______

11. $\frac{21}{7} =$ ______

12. $\frac{40}{8} =$ ______

13. $49 \div 7 =$ _____

14. $32 \div 8 =$ _____

15. $\frac{64}{8} =$ _____

16. $\frac{56}{7} =$ _____

Find the area.

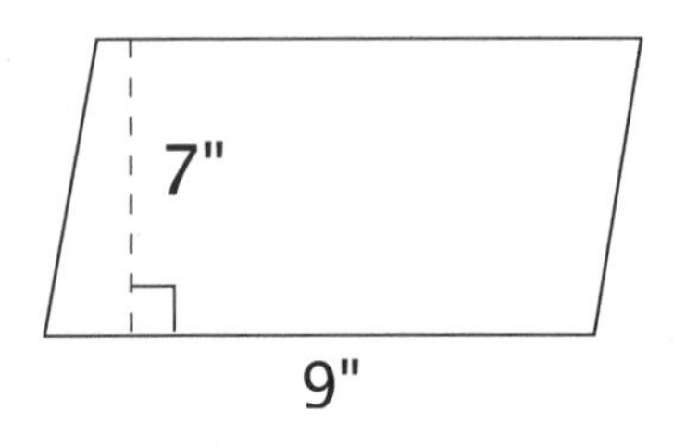

17. A = ______

18. Which bag of apples weighs more, the three-pound bag or the 50-ounce bag? _____

19. Aiden gave me 28 quarters. How many dollars did I receive?

_____

20. Find the average of the numbers: 1, 2, 7, 10, 12, 16 _____

UNIT TEST **Lessons 7-12**

II

Divide.

1. $4\overline{)36}$

2. $6\overline{)24}$

3. $8\overline{)32}$

4. $7\overline{)49}$

5. $8\overline{)56}$

6. $4\overline{)20}$

7. $6\overline{)42}$

8. $6\overline{)60}$

9. $16 \div 8 =$ ______

10. $35 \div 7 =$ ______

11. $\frac{48}{8} =$ ______

12. $\frac{7}{7} =$ ______

13. $12 \div 4 =$ ______

14. $36 \div 6 =$ ______

Divide.

15. $\frac{70}{7} =$ ______

16. $\frac{12}{6} =$ ______

17. $8\overline{)8}$

18. $7\overline{)56}$

19. $4\overline{)28}$

20. $7\overline{)42}$

21. $8\overline{)24}$

22. $4\overline{)32}$

23. $8\overline{)80}$

24. $7\overline{)21}$

25. $18 \div 6 =$ ______

26. $4 \div 4 =$ ______

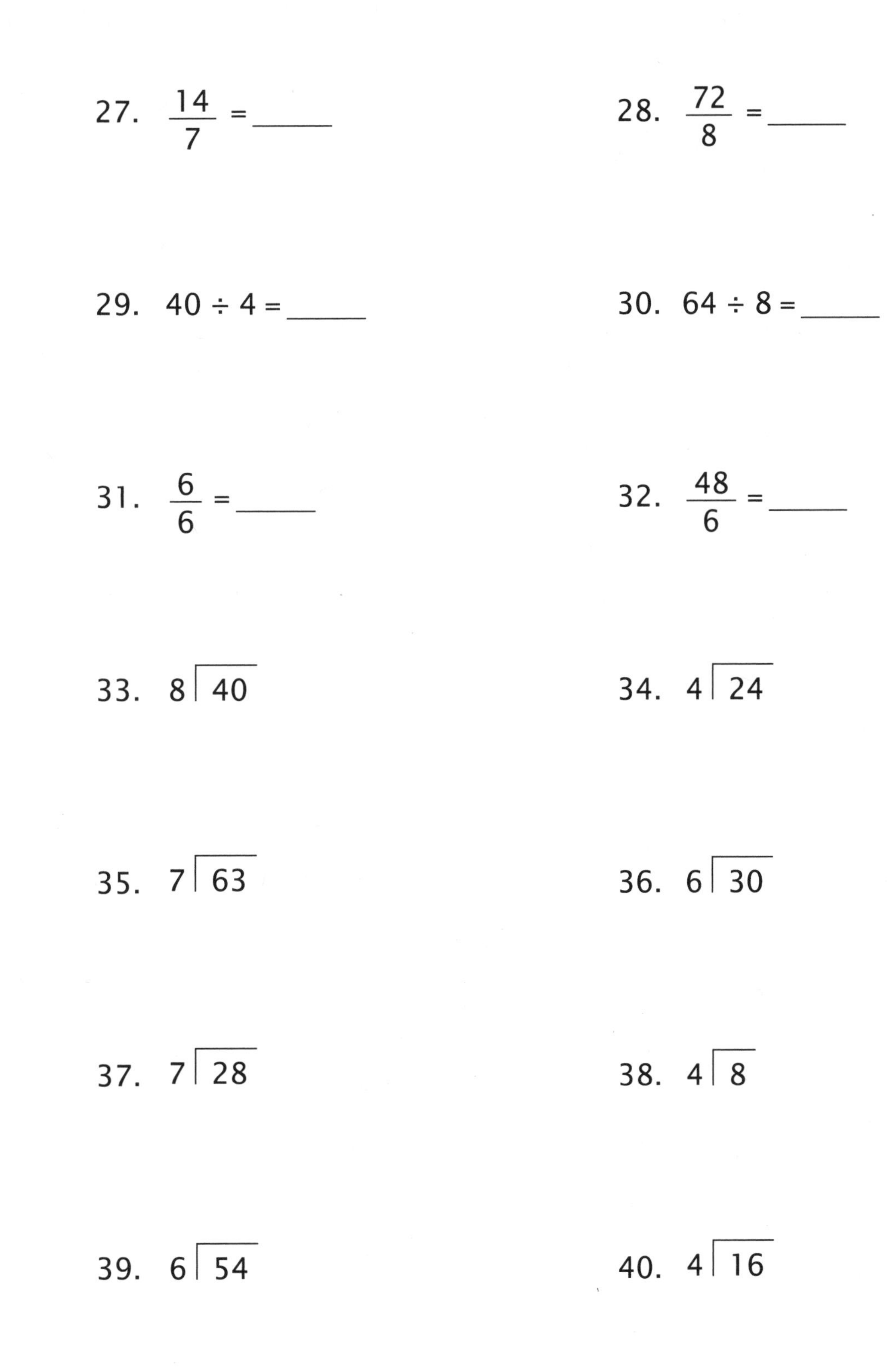

27. $\frac{14}{7}$ = ______

28. $\frac{72}{8}$ = ______

29. $40 \div 4 =$ ______

30. $64 \div 8 =$ ______

31. $\frac{6}{6}$ = ______

32. $\frac{48}{6}$ = ______

33. $8\overline{)40}$

34. $4\overline{)24}$

35. $7\overline{)63}$

36. $6\overline{)30}$

37. $7\overline{)28}$

38. $4\overline{)8}$

39. $6\overline{)54}$

40. $4\overline{)16}$

Follow the signs.

41. $\begin{array}{r} 35 \\ 72 \\ 15 \\ +48 \\ \hline \end{array}$

42. $\begin{array}{r} 91 \\ -36 \\ \hline \end{array}$

43. $\begin{array}{r} 75 \\ \times 58 \\ \hline \end{array}$

44. What is the area of a triangle with a base of seven yards and a height of two yards? _____

45. Find the average of the numbers: 5, 9, 13 _____

46. How many quarts are there in six gallons? _____

47. How many gallons are there in 32 quarts? _____

48. How many quarters are there in $9? _____

49. How many dollars are equal to 20 quarters? _____

50. How many ounces are in two pounds? _____

Find the area of the figures.

1.

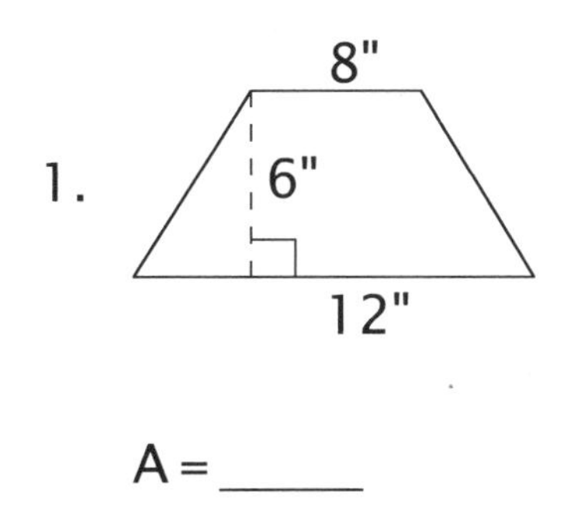


A = ______

2.

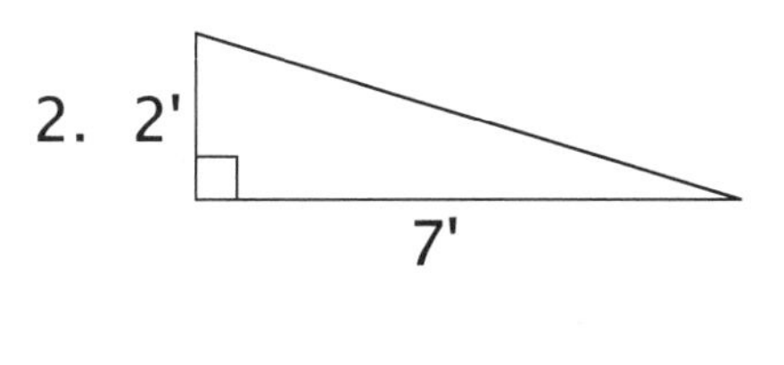

A = ______

3.

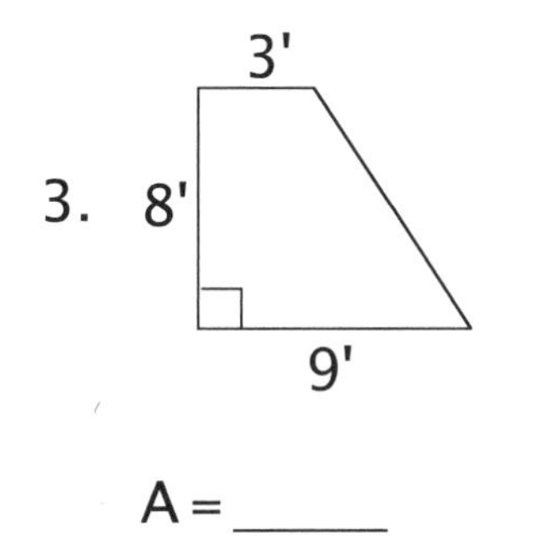


A = ______

4.

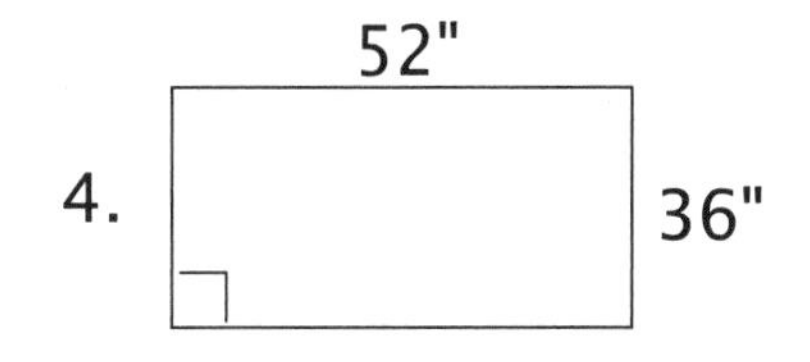


A = ______

Multiply using the shortcut method.

5. 14 x 20 = ______

6. 22 x 30 = ______

7. 43 x 30 = ______

8. 51 x 20 = ______

Divide.

9. 42 ÷ 7 = _____

10. 100 ÷ 10 = _____

11. 64 ÷ 8 = _____

12. 27 ÷ 9 = _____

Add or subtract.

13. $\begin{array}{r} 352 \\ +\ 126 \\ \hline \end{array}$

14. $\begin{array}{r} 811 \\ -\ 349 \\ \hline \end{array}$

15. $\begin{array}{r} 607 \\ +\ 785 \\ \hline \end{array}$

Fill in the blanks.

16. 8 qt = _____ gal

17. 28 quarters = _____ dollars

18. 27 ft = _____ yd

19. A 50-foot row of corn needs to be weeded. Five people each weeded the same amount. How many feet of corn did each person weed? _____

20. Jordan needs $315 to buy the bicycle he wants. If he has saved $227, how much more does he need to save? _____

Follow the directions.

1. Write using numbers: 200,000 + 20,000 + 1,000 + 300 + 40 + 6

_______________

2. Write using numbers: 3,000,000 + 400,000 + 60,000 + 7,000

_______________

3. Write using place-value notation: 6,123,500

_______________

4. Write using place-value notation: 4,500,000

_______________

Find the area of the figures.

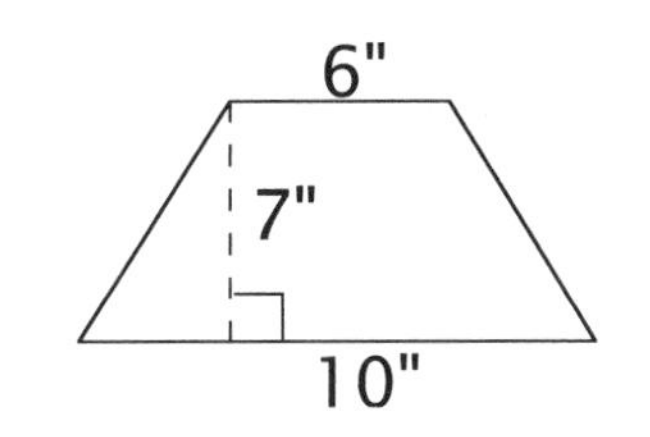


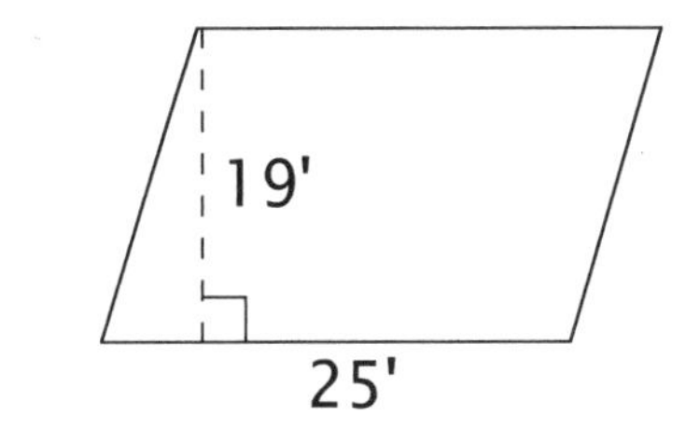


5. A = _____

6. A = _____

Multiply.

7. 6 x 200 = _____

8. 24 x 200 = _____

9. 17 x 100 = _____

10. 32 x 200 = _____

Divide.

11. 56 ÷ 8 = ____

12. 49 ÷ 7 = ____

13. 9 ÷ 9 = ____

14. 24 ÷ 6 = ____

15. How many parallel lines does a trapezoid have? _____

16. How many ounces are there in a five-pound bag of flour? _____

17. How many square inches are there in a triangle with a base of eight inches and a height of two inches? _____

18. Chloe bought 16 pints of ice cream. How many quarts of ice cream does she have? _____

How many gallons of ice cream does she have? _____

Follow the directions.

1. Write in standard notation and read the number:

   7 x 1,000,000,000 + 6 x 100,000,000 + 3 x 10,000,000 +
   2 x 1,000,000 + 4 x 100,000 ____________________

2. Write in standard notation and read the number:

   5 x 100,000,000 + 5 x 10,000,000 + 5 x 1,000,000 +
   4 x 100,000 + 3 x 10,000 + 1 x 1,000 ____________________

3. Write in standard notation:

   one trillion, six hundred thirty-five billion,
   seven hundred twenty-one million ____________________

4. Write in standard notation:

   four million, three hundred fifteen thousand, twenty-one

   ____________________

5. Write using expanded notation: 8,250,000,000

   ________________________________________

6. Write using expanded notation: 3,400,000,000,274

   ________________________________________

Multiply.

7. $10 \times 60 =$ _____

8. $12 \times 40 =$ _____

9. $20 \times 200 =$ _____

Add or subtract.

10.
$$\begin{array}{r} 2{,}543 \\ +\ 8{,}067 \\ \hline \end{array}$$

11.
$$\begin{array}{r} 6{,}460 \\ -\ 192 \\ \hline \end{array}$$

12.
$$\begin{array}{r} 1{,}247 \\ 3{,}598 \\ +\ 6{,}013 \\ \hline \end{array}$$

Divide.

13. $63 \div 9 =$ _____

14. $36 \div 6 =$ _____

15. $21 \div 7 =$ _____

16. $8 \div 8 =$ _____

17. Find the average of the numbers: 3, 5, 11, 13 _____

18. Mackenzie spent 16 hours a month babysitting. How many hours did she babysit in 12 months? _____

Divide.

1. $3\overline{)10}$

2. $6\overline{)25}$

3. $9\overline{)30}$

4. $4\overline{)31}$

5. $2\overline{)15}$

6. $5\overline{)47}$

7. $8\overline{)20}$

8. $7\overline{)41}$

9. $8\overline{)37}$

Add or subtract.

10. $\begin{array}{r} 3,076 \\ -1,467 \\ \hline \end{array}$

11. $\begin{array}{r} 4,654 \\ -3,298 \\ \hline \end{array}$

12. $\begin{array}{r} 6,512 \\ +7,285 \\ \hline \end{array}$

Multiply.

13. 11 x 700 = _____

14. 12 x 300 = _____

15. 21 x 400 = _____

16. Write using standard notation:

eight billion, three hundred ten million, six hundred seventy-five thousand, four hundred twenty

_______________________

17. A pet-shop owner has 18 canaries. If he put four birds in a cage, how many full cages will he have? _____

How many birds will be left over for another cage? _____

18. What is the area of a rectangle with a base of 45 feet and a height of 38 feet? _____

Rewrite each problem using place-value notation, and multiply. Compare your answers.

1. $\begin{array}{r} 252 \\ \times\ 4 \\ \hline \end{array}$ $\quad \times$ ______

2. $\begin{array}{r} 4 \\ \times\ 252 \\ \hline \end{array}$ $\quad \times$ ______

Divide.

3. $3\overline{)90}$

4. $6\overline{)240}$

5. $2\overline{)120}$

6. $5\overline{)200}$

7. $6\overline{)55}$

8. $8\overline{)26}$

9. $9\overline{)64}$

10. $7\overline{)30}$

Fill in the blanks.

11. 3 lb = ___________ oz

12. 4 gal = ___________ qt

13. 33 ft = ___________ yd

14. 1 ton = ___________ lb

15. 6 tons = ___________ lb

16. 3 tons = ___________ lb

17. Jacob wants to divide 31 dollar bills among his four children. How many will each child receive? _____

How many dollar bills will Jacob have left over? _____

18. Write in standard notation:

five trillion, four hundred billion, six hundred million, five

________________________

Divide. Check your work by multiplying upside down.

1. $2\overline{)42}$

2. $9\overline{)92}$

3. $3\overline{)67}$

4. $4\overline{)19}$

5. $5\overline{)21}$

6. $8\overline{)480}$

Write the numbers in columns and add.

7. 25 + 75 + 45 = _____

8. 8 + 7 + 4 + 3 + 2 = _____

9. 95 + 345 =_____

Fill in the blanks.

10. 4 tons = __________ lb

11. 31 dollars = ______ quarters

12. 10 lb = __________ oz

13. 1 mile = __________ feet

14. 3 mi = __________ ft

15. 5 mi = __________ ft

16. Find the average of the numbers: 4, 8, 15, 25 _____

17. Taylor has 25 pictures to put in his photo album. He can fit four pictures on a page. How many pages will he fill? _____

    How many pictures will he have left over? _____

18. How many feet are there in five yards? _____

Divide. Check your work by multiplying.

1. $3\overline{)714}$

2. check for #1

3. $5\overline{)628}$

4. check for #3

5. $4\overline{)368}$

6. check for #5

7. $7\overline{)113}$

8. check for #7

Multiply.

9. $\begin{array}{r} 453 \\ \times 46 \\ \hline \end{array}$

10. $\begin{array}{r} 839 \\ \times 25 \\ \hline \end{array}$

11. $\begin{array}{r} 851 \\ \times 69 \\ \hline \end{array}$

Fill in the blanks.

12. 40 yd = __________ ft

13. 2 mi = __________ ft

14. 5 lb = __________ oz

15. Write in standard notation:

2 x 1,000,000,000 + 4 x 100,000,000 + 9 x 10,000,000 +
5 x 1,000,000

______________________

16. Are the lines shown parallel or perpendicular? ______________

Divide. Write all remainders as fractions, and check your work by multiplying.

1. $5\overline{)895}$

2. check for #1

3. $8\overline{)356}$

4. check for #3

5. $3\overline{)614}$

6. check for #5

7. $6\overline{)578}$

8. check for #7

9. Driving at a steady speed, Emma covered 300 miles in 5 hours. How many miles did Emma drive each hour? _____

10. How many feet are there in three miles? _____

11. The area of a parallelogram is 35 square yards. If the base is seven yards, what is the height? _____

12. Change the base and height of the parallelogram in #11 to feet, and then multiply to find the area in square feet. _____

13. Which weighs more, eight tons or 1,600 pounds? _____

14. How many perpendicular corners does a rectangle have? _____

15. Richard was born in 1974. How old was he on his birthday in 2003? _____

Fill in the blanks.

1. 29 to the nearest ten is _____.

2. 109 to the nearest hundred is _____.

3. 1,168 to the nearest thousand is _____.

4. 14 to the nearest ten is _____.

5. 355 to the nearest hundred is _____.

6. 4,500 to the nearest thousand is _____.

Estimate the answer, and then divide and compare your answers.

7. $6\overline{)869} \rightarrow 6\overline{)(\quad\quad)}$

8. $6\overline{)869}$

Divide and check by multiplying.

9. $2\overline{)652}$

10. check for #9

11. $7\overline{)239}$

12. check for #11

13. Abigail walked two miles. How many feet did she walk? _____

14. How many pounds are there in five tons? _____

15. Thomas drove 500 miles every day for five days. How many miles did he drive in all? _____

16. What is the area of a trapezoid with bases of 13 and 17 feet and a height of 11 feet? _____

Find the area of the trapezoids.

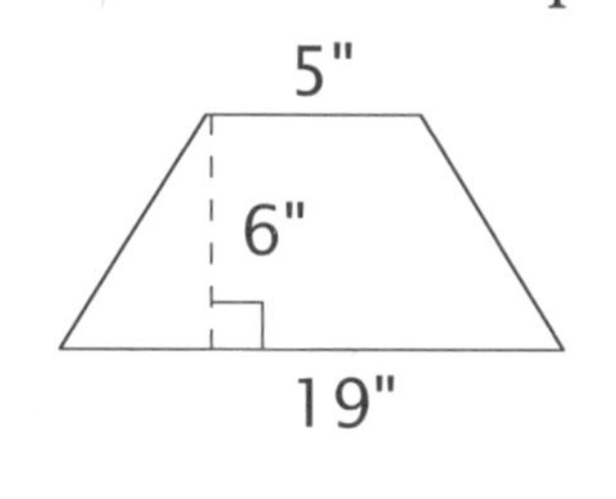


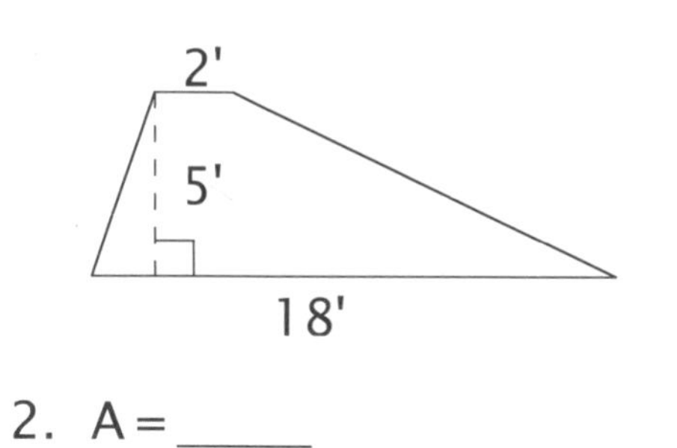


1. A = _____

2. A = _____

Follow the directions.

3. Write in standard notation:

   3 x 1,000,000,000 + 7 x 100,000,000 + 6 x 10,000,000 + 1 x 1,000,000 + 8 x 100,000

   ______________________________

4. Write in standard notation:

   two trillion, four hundred thirteen billion, two hundred eighty-three million, three hundred fifty thousand

   ______________________________

5. Write using expanded notation: 75,123,000

   ______________________________

Multiply.

6. 28 x 60 = _____

7. 13 x 400 = _____

8. 56 x 700 = _____

Divide. Write your remainders without using fractions.

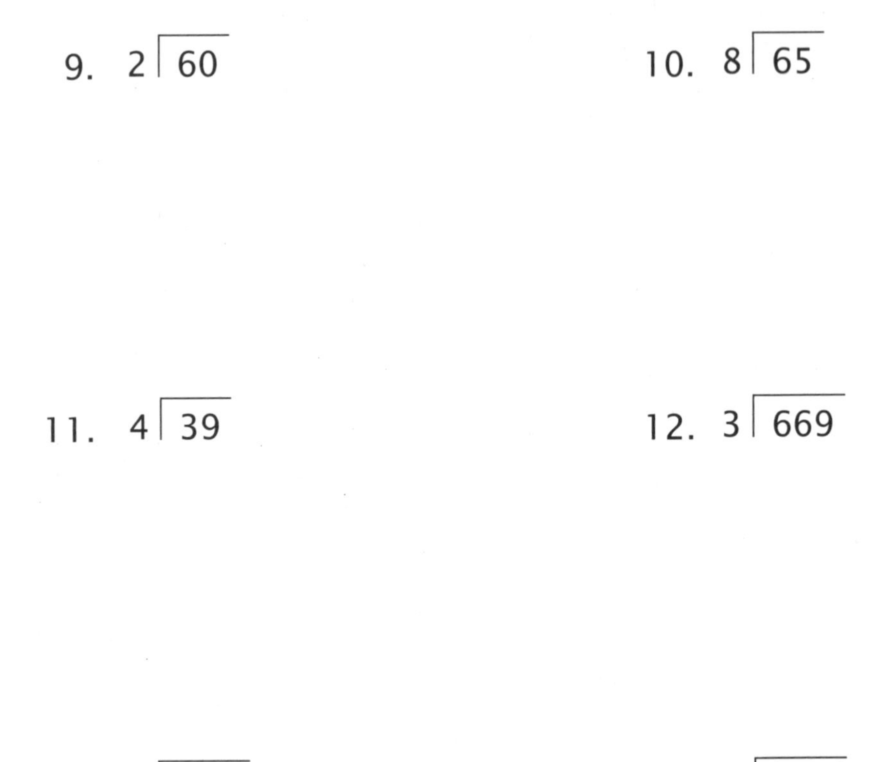

9. $2\overline{)60}$

10. $8\overline{)65}$

11. $4\overline{)39}$

12. $3\overline{)669}$

13. $5\overline{)257}$

14. $8\overline{)829}$

Fill in the blanks.

15. 31 to the nearest ten is _____ .

16. 249 to the nearest hundred is _____ .

17. 2,503 to the nearest thousand is _____ .

Estimate the answer, and then divide and compare your answers. Write the remainders as fractions.

18. $5\overline{)908} \rightarrow 5\overline{)(\quad\quad)}$

19. $5\overline{)908}$

Divide and check by multiplying. Write the remainders as fractions.

20. $4\overline{)865}$

21. check for #20

22. $2\overline{)157}$

23. check for #22

24. How many feet are there in five miles? _____

25. How many pounds are there in nine tons? _____

26. Five hundred and sixty-eight ears listened to Mr. Demme's speech. How many people were listening? _____

Divide, and then check by multiplying. Write any remainders as fractions.

1. $34\overline{)517}$

2. check for #1

3. $18\overline{)367}$

4. check for #3

5. $5\overline{)78}$

6. check for #5

7. $9\overline{)934}$

8. check for #7

Find the area of the figures.

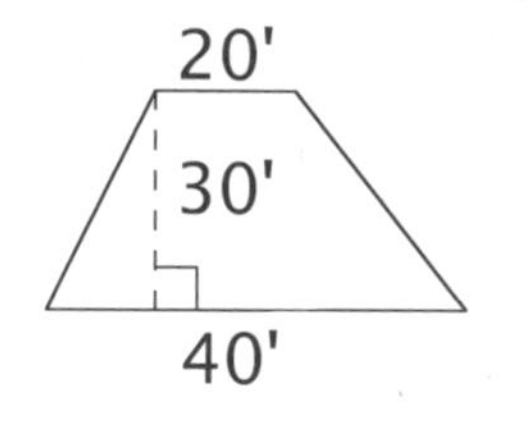


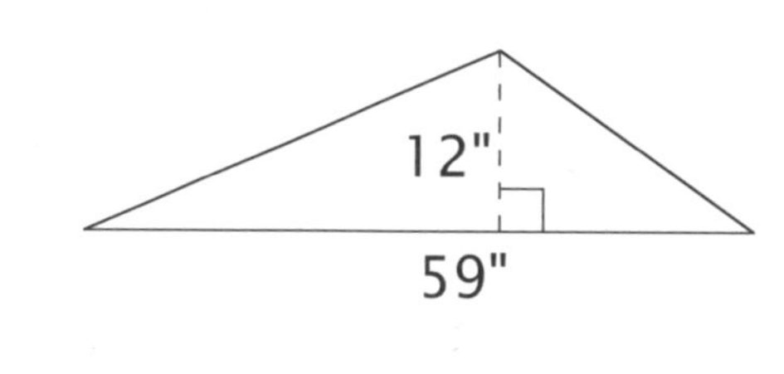


9. ______

10. ______

Find the area of the figure.

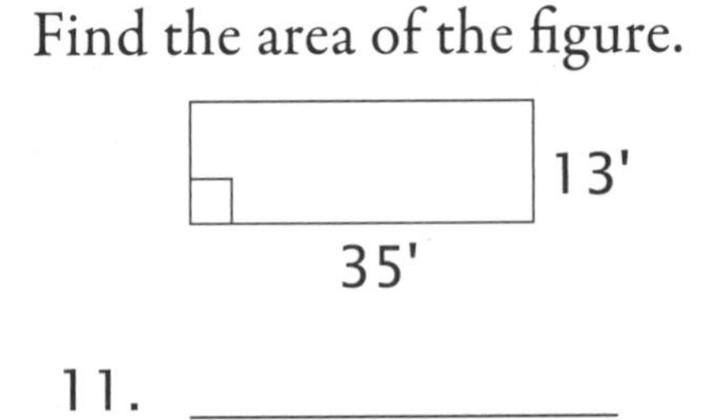


11. ______________

Fill in the blanks.

12. 144 in = _____ ft

13. 20 feet = _____ inches

14. 72 in = _____ ft

15. Jaden has 48 quarters. How many dollars is that? _____

16. Logan drove at 55 miles an hour for 330 miles. Divide by 55 to find how many hours the trip took. _____

17. On Logan's 330-mile trip, he used 11 gallons of gasoline. How many miles did he go for each gallon? _____

18. Which is greater, 7 tons or 140,000 pounds? _____

Divide, and then check by multiplying. Write any remainders as fractions.

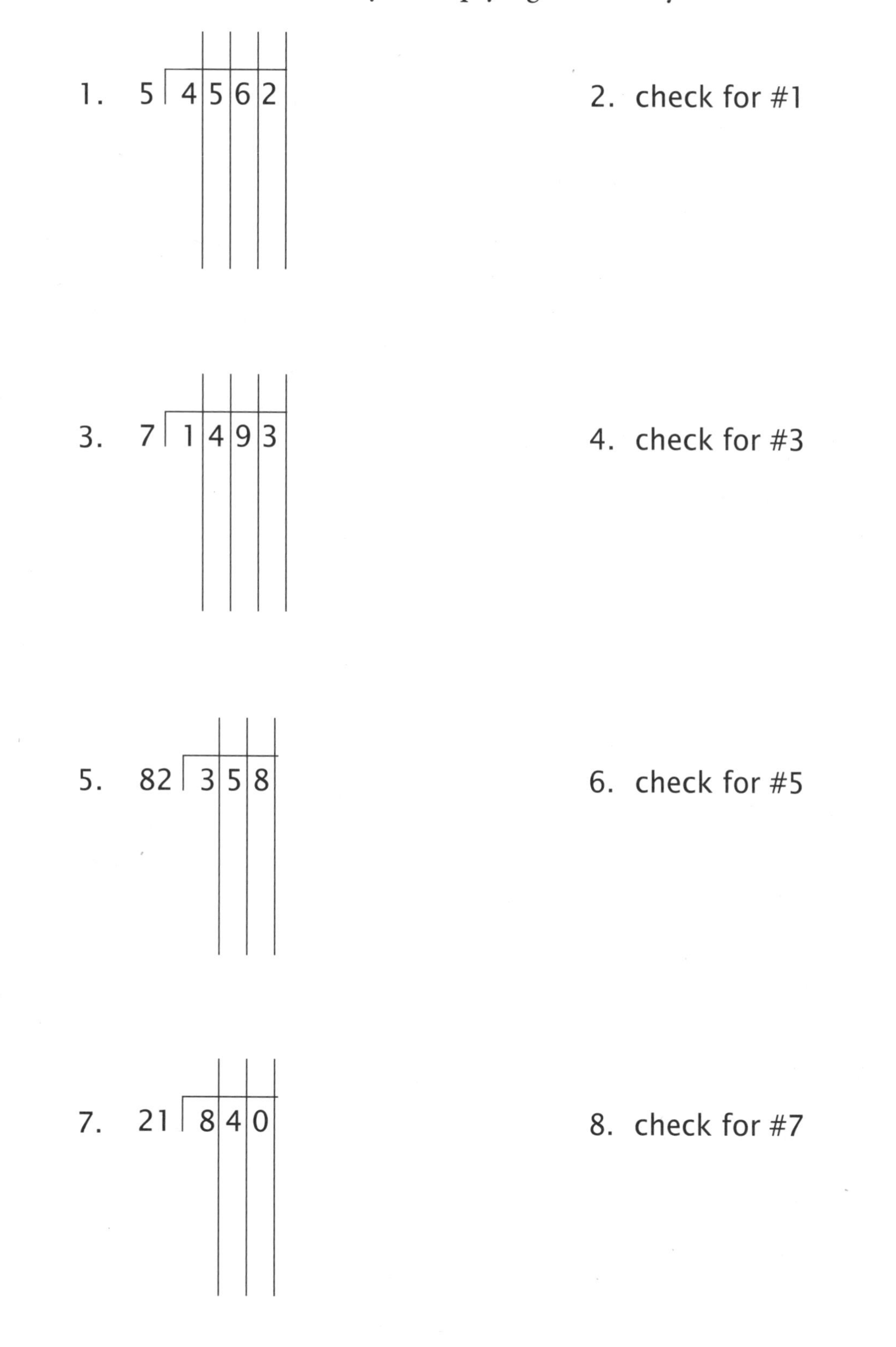

Multiply.

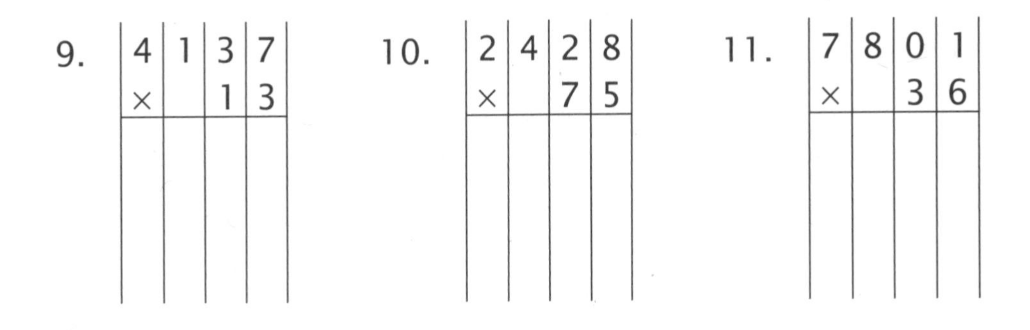


12. How many feet are there in 396 inches? _____

13. How many parallel lines are found in an uppercase "E"? _____

14. Cameron wants to hang a picture that weighs 80 ounces. The label on the package of the picture hanger he plans to buy says the hanger will hold 10 pounds. Is the hanger strong enough for his picture? _____

15. A pool of water held 2,272 gallons. How many quarts of water did the pool hold? _____

Divide, and then check by multiplying. Write any remainders as fractions.

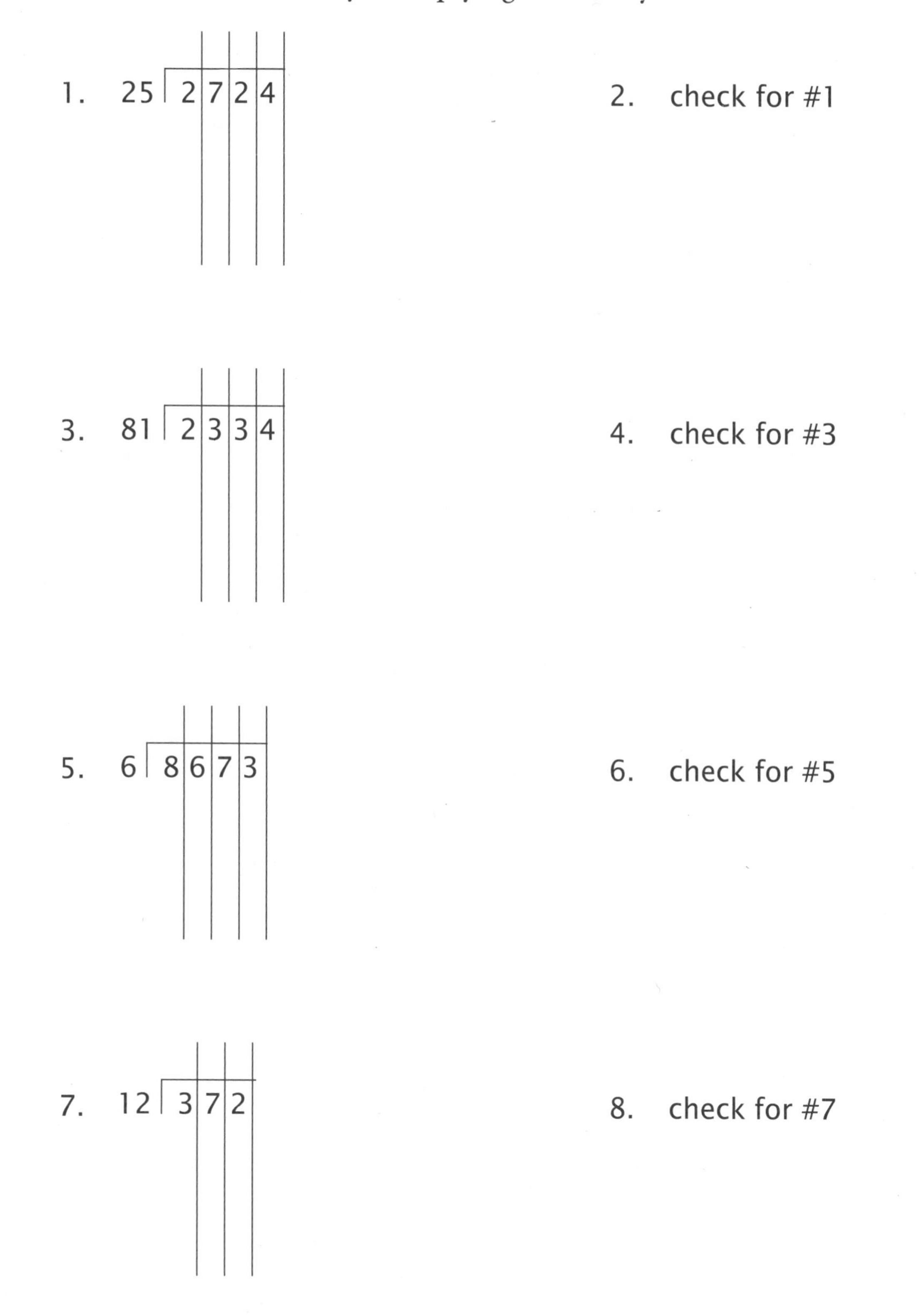

Multiply upside down.

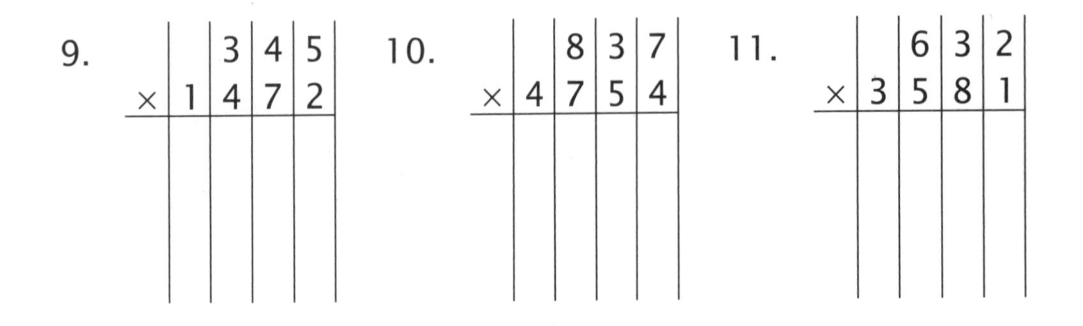

9. 345 × 1472

10. 837 × 4754

11. 632 × 3581

12. What is the area of a trapezoid with bases of 13 feet and 19 feet and a height of 41 feet? _____

13. Thirty-four people want to go to the picnic. If all the available cars hold five people each, how many cars are needed to take everyone who wants to go? _____

14. Draw two lines that look parallel.

Divide, and then check by multiplying. Write any remainders as fractions.

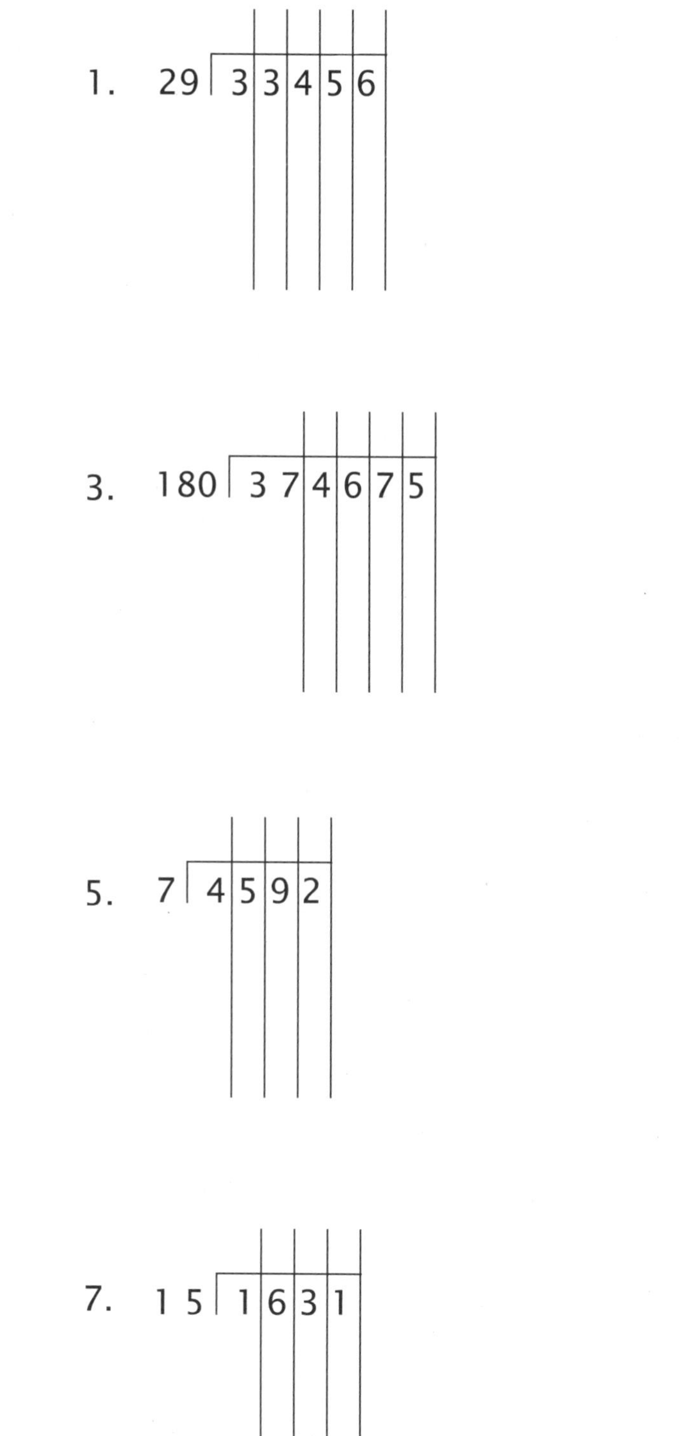


2. check for #1

4. check for #3

6. check for #5

8. check for #7

Find the missing part of each parallelogram or rectangle.

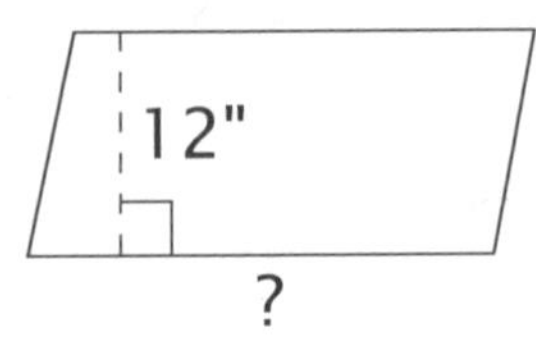


?

250'

9. area = 336 sq in
   base = _____

10. area = 37,500 sq ft
    height = _____

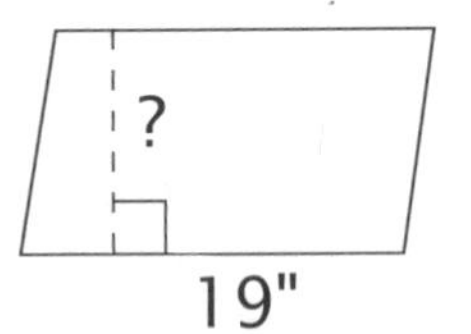


11. area = 323 sq in
    height = _____

12. Riley weighs 56 pounds. How many ounces does she weigh?

    _____

13. Tyler drove 270 miles in 6 hours. How many miles an hour was he driving? _____

14. Find the average weight:
    780 pounds; 4,670 pounds; 550 pounds. _______________

    What is the average weight in tons? _______________

Find the volume.

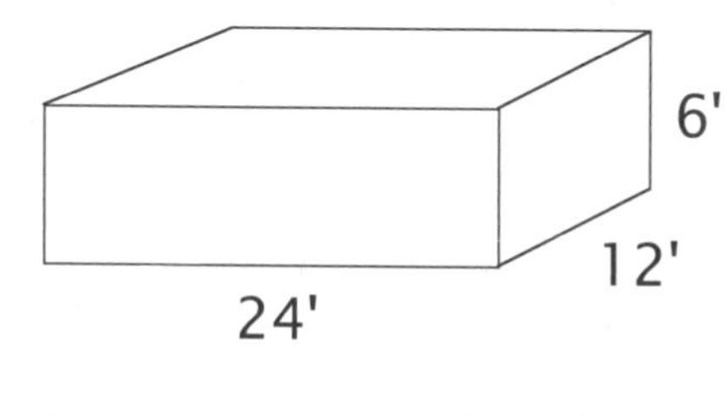


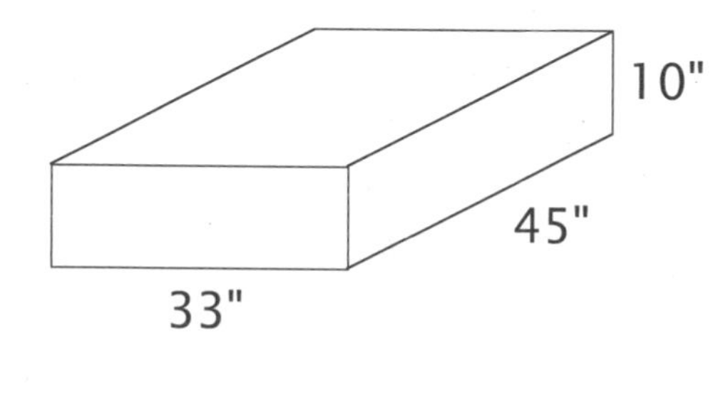


1. V = ________ cubic feet

2. V = ________ cubic inches

Divide, and then check by multiplying. Use estimation to help you if needed.

3.

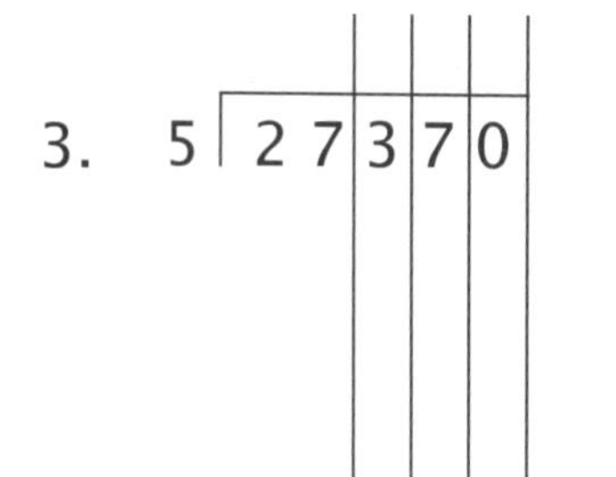


4. check for #3

5.

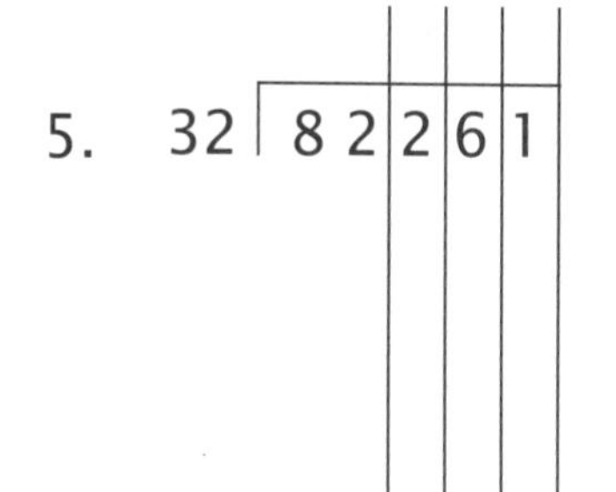


6. check for #5

Fill in the blanks.

7. 28 pt = __________ qt

8. 44 dollars = ______ quarters

9. 48 oz = __________ lb

10. 6 tons = __________ lb

11. 100 qt = __________ gal

12. 60 in = __________ ft

13. A rectangular pool measures 20 feet by 15 feet by 6 feet deep. How many cubic feet of water would fill the pool to the brim?

_____

14. How many gallons of water are needed to fill the pool in #13?

_____

15. At eight pounds per gallon, what will the water in the pool in #13 weigh when the pool is filled to the brim? _____

Solve.

1. $\frac{4}{7}$ of 21 = _____

2. $\frac{2}{3}$ of 9 = _____

3. $\frac{1}{9}$ of 36 = _____

4. $\frac{3}{5}$ of 75 = _____

5. $\frac{5}{6}$ of 18 = _____

6. $\frac{1}{2}$ of 28 = _____

Divide, and then check by multiplying. Use estimation to help you if needed.

7. 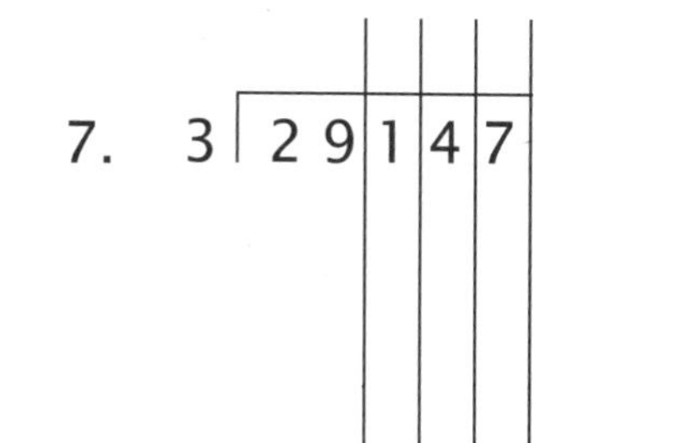


8. check for #7

9. 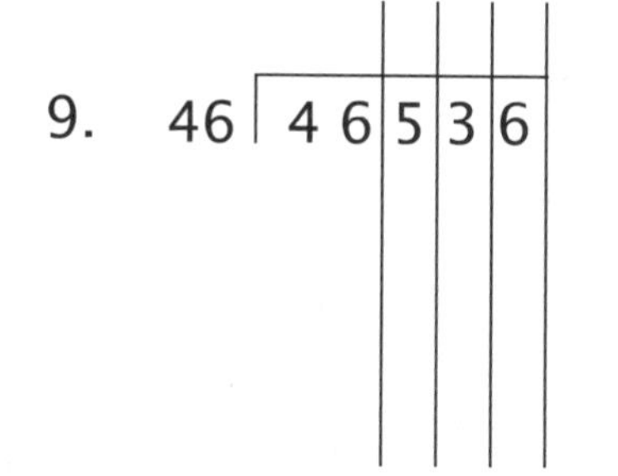


10. check for #9

Fill in the blanks.

11. 36 ft = _____ yd

12. 24 quarters = _____ dollars

13. 8 mi = __________ ft

14. Isabella collected 32 flowers for a nature project. She decided to press and keep 1/4 of them. How many flowers did she keep?

__________________________

15. What is the volume of a rectangular shape that measures 100 inches by 100 inches by 90 inches?

__________________________

Write the number represented by the Roman numerals.

1. XIV = ______

2. LXXII = ______

3. CCXXX = ______

4. XCIX = ______

Use Roman numerals to represent the number.

5. 41 = ______

6. 85 = ______

7. 333 = ______

8. 29 = ______

Solve.

9. $\frac{1}{2}$ of 12 = ______

10. $\frac{3}{8}$ of 64 = ______

11. $\frac{2}{5}$ of 15 = ______

Find the area.

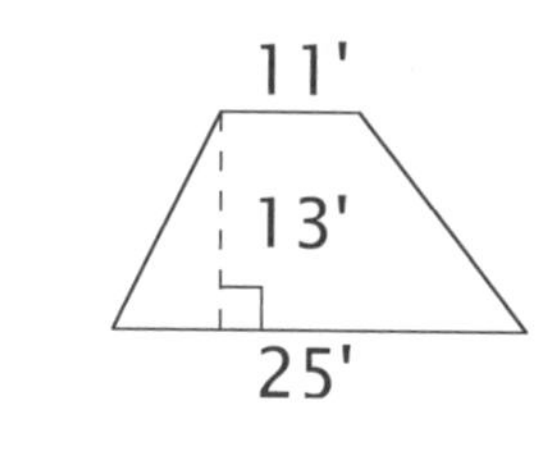


12. A = ______

Divide, and then check by multiplying. Use estimation to help you if needed.

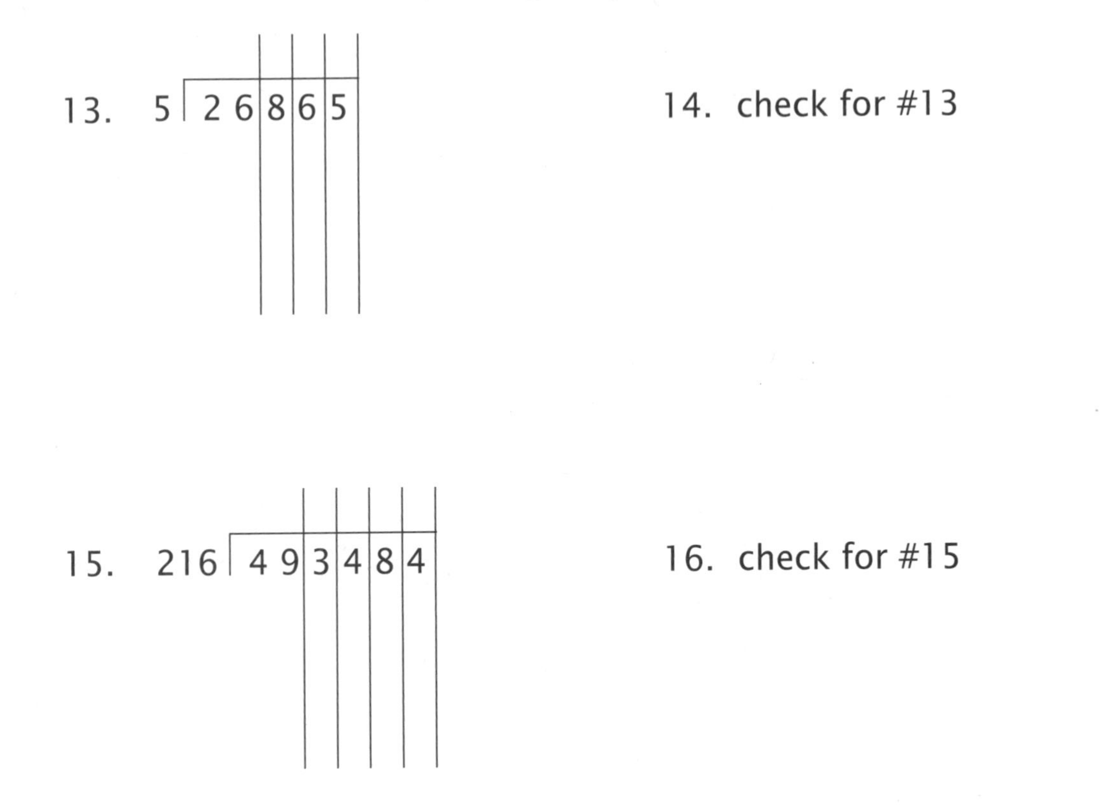


17. Find the average of the numbers: 16, 43, 58, 91 _____

18. Morgan spotted 24 horses grazing in a field. One-sixth of them were black. How many of the horses were black? _____

Find the denominators and numerators of the rectangles.

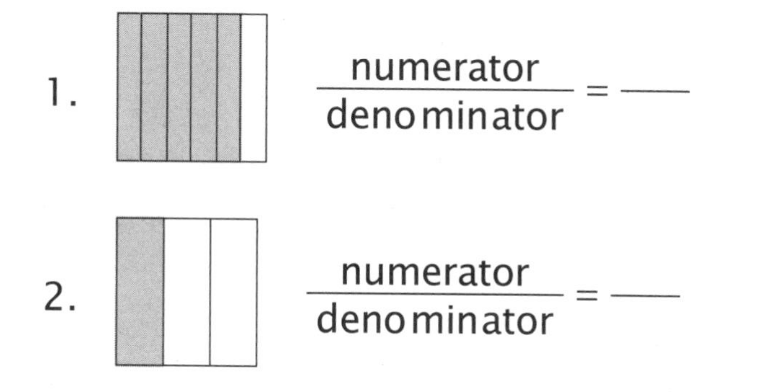


Shade the rectangles to show the given fractions.

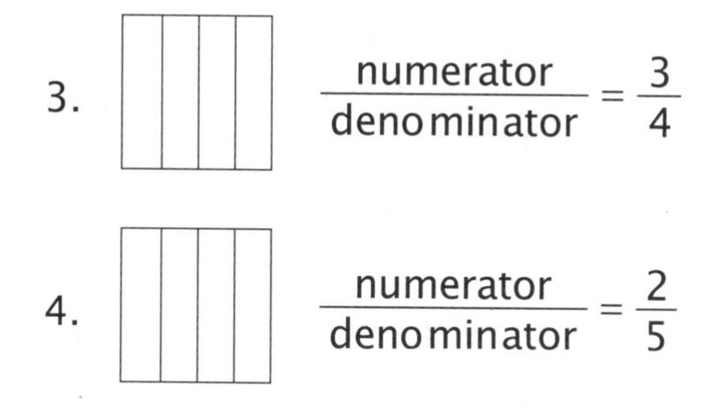


Write the number represented by the Roman numerals.

5. XXVI = _____

6. XLIII = _____

7. CLXV = _____

8. CXCII = _____

Use Roman numerals to represent the number.

9. 47 = _____

10. 18 = _____

11. 219 = _____

12. 154 = _____

Divide, and then check by multiplying. Use estimation to help you if needed.

13. $31 \overline{)8793}$

14. check for #13

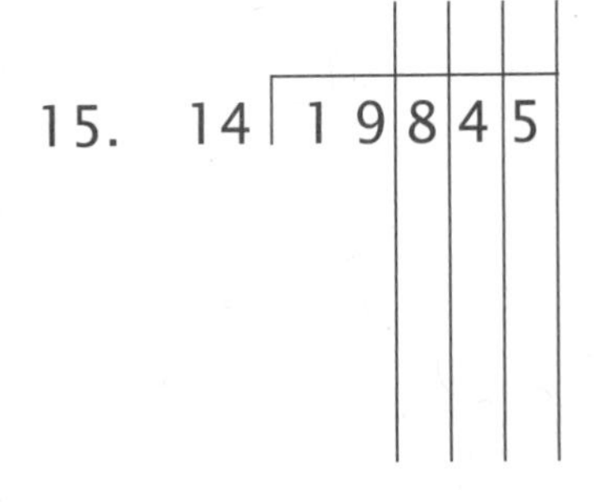

15. $14 \overline{)19845}$

16. check for #15

17. What is the area of a triangle with a base of 20 inches and a height of 9 inches? _____

18. Paige drove 165 miles at 55 miles an hour. How many hours did the trip take? _____

Write the number represented by the Roman numerals.

1. MMCC = _____

2. DXXV = _____

3. DCCL = _____

4. CMXXIX = _____

Use Roman numerals to represent the number.

5. 58 = _____

6. 520 = _____

7. 3,700 = _____

8. 1965 = _____

Find the denominators and numerators of the rectangles.

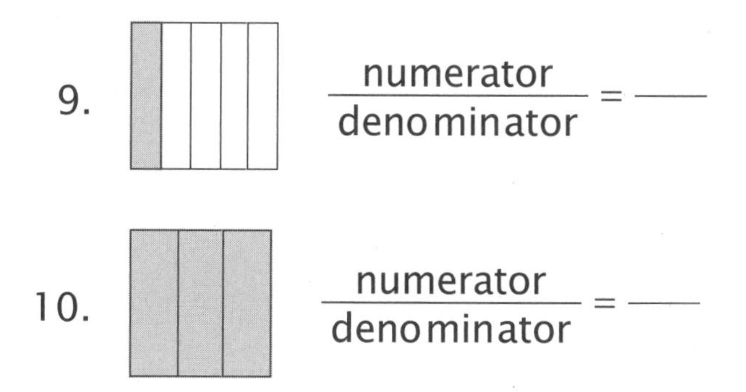

Divide, and then check by multiplying. Use estimation to help you if needed.

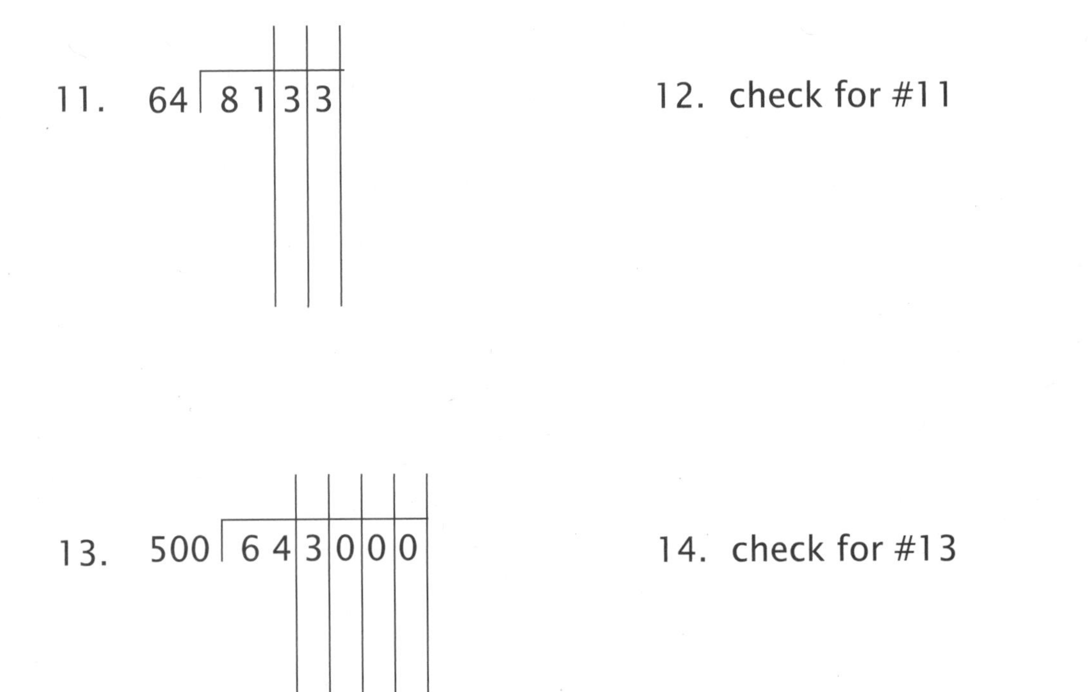


15. What is the area of a trapezoid with bases of seven feet and nine feet and a height of eight feet? _____

16. Round 561 to the nearest hundred. _____

17. At an average speed of 15 miles an hour, how long would it take to travel 135 miles? _____

18. Mom bought 25 pounds of apples. One-fifth of the apples spoiled before she could use them. How many pounds of good apples does she have left? _____

Divide. Write your remainders with fractions. Check your answers by multiplying.

1. $13\overline{)568}$

2. check for #1

3. $30\overline{)971}$

4. check for #3

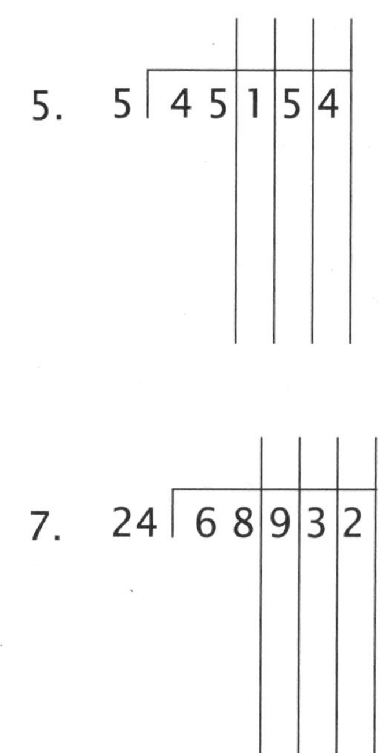


6. check for #5

8. check for #7

Find the volume.

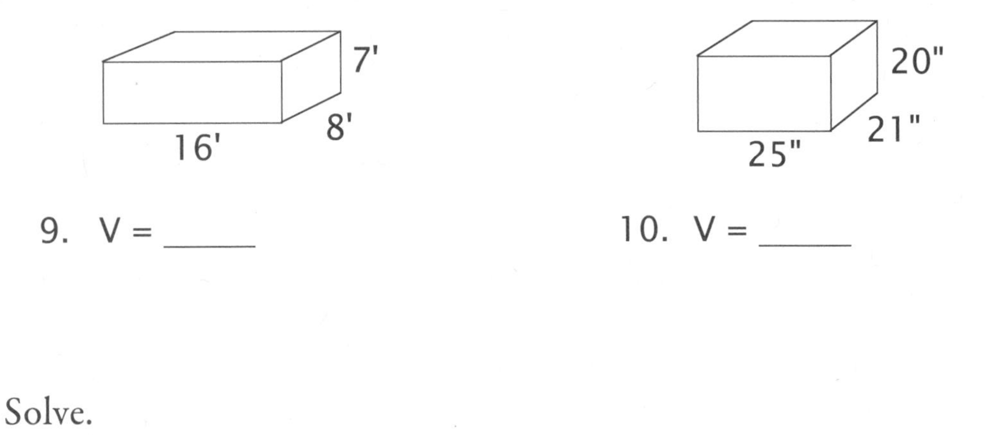


9. V = _____

10. V = _____

Solve.

11. $\frac{1}{2}$ of 32 = _____

12. $\frac{5}{8}$ of 64 = _____

13. $\frac{5}{6}$ of 42 = _____

Find the denominators and numerators of the rectangles.

14. $\frac{\text{numerator}}{\text{denominator}}$ = ___

15. $\frac{\text{numerator}}{\text{denominator}}$ = ___

16. How many inches long is an eight-foot board? _____

17. What number is represented by the Roman numeral XCIX ? _____

18. Write the given number with Roman numerals: 2,453 _____

Divide. Write your remainders without using fractions.

1. $4\overline{)80}$

2. $7\overline{)53}$

3. $8\overline{)648}$

4. $5\overline{)396}$

Divide. Write your remainders with fractions. Check your answers by multiplying.

5. $25\overline{)631}$

6. check for #5

7. $16\overline{)349}$

8. check for #7

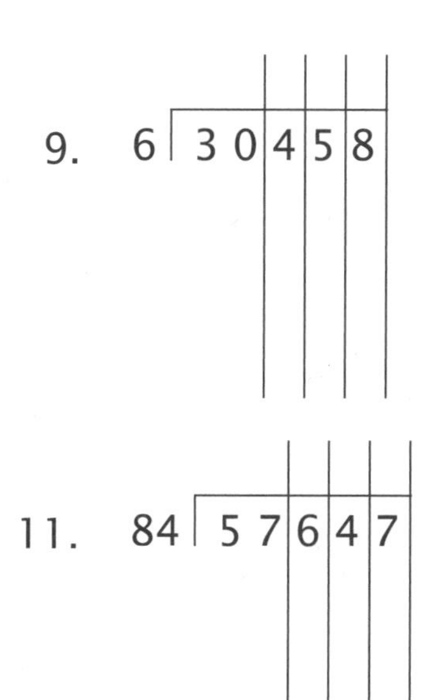

9. $6\overline{)30458}$

10. check for #9

11. $84\overline{)57647}$

12. check for #11

Find the area of each figure.

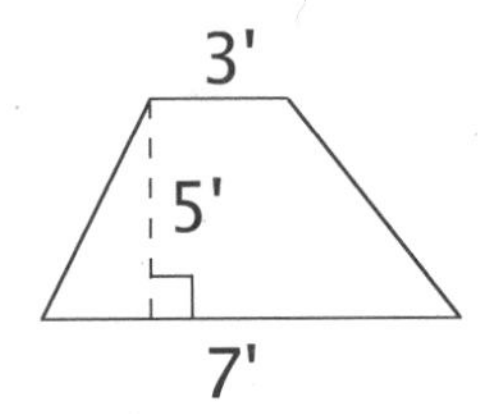


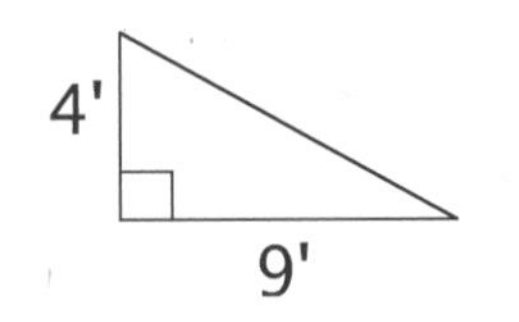


13. A = ______

14. A = ______

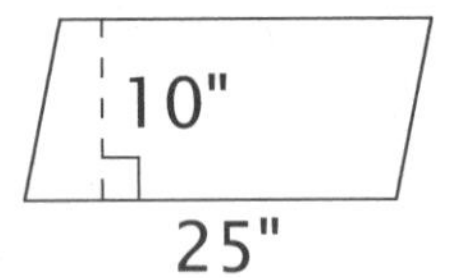


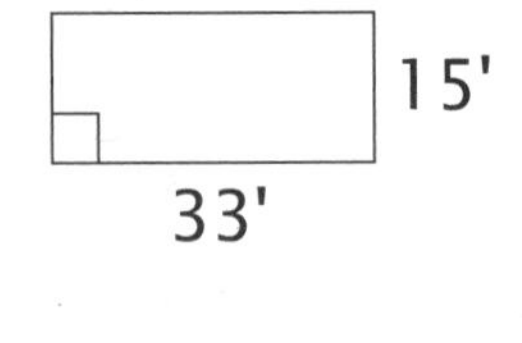


15. A = ______

16. A = ______

Find the volume.

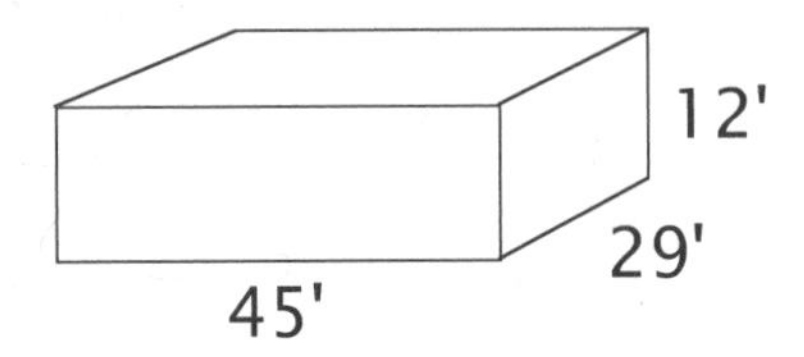


17. V = ______

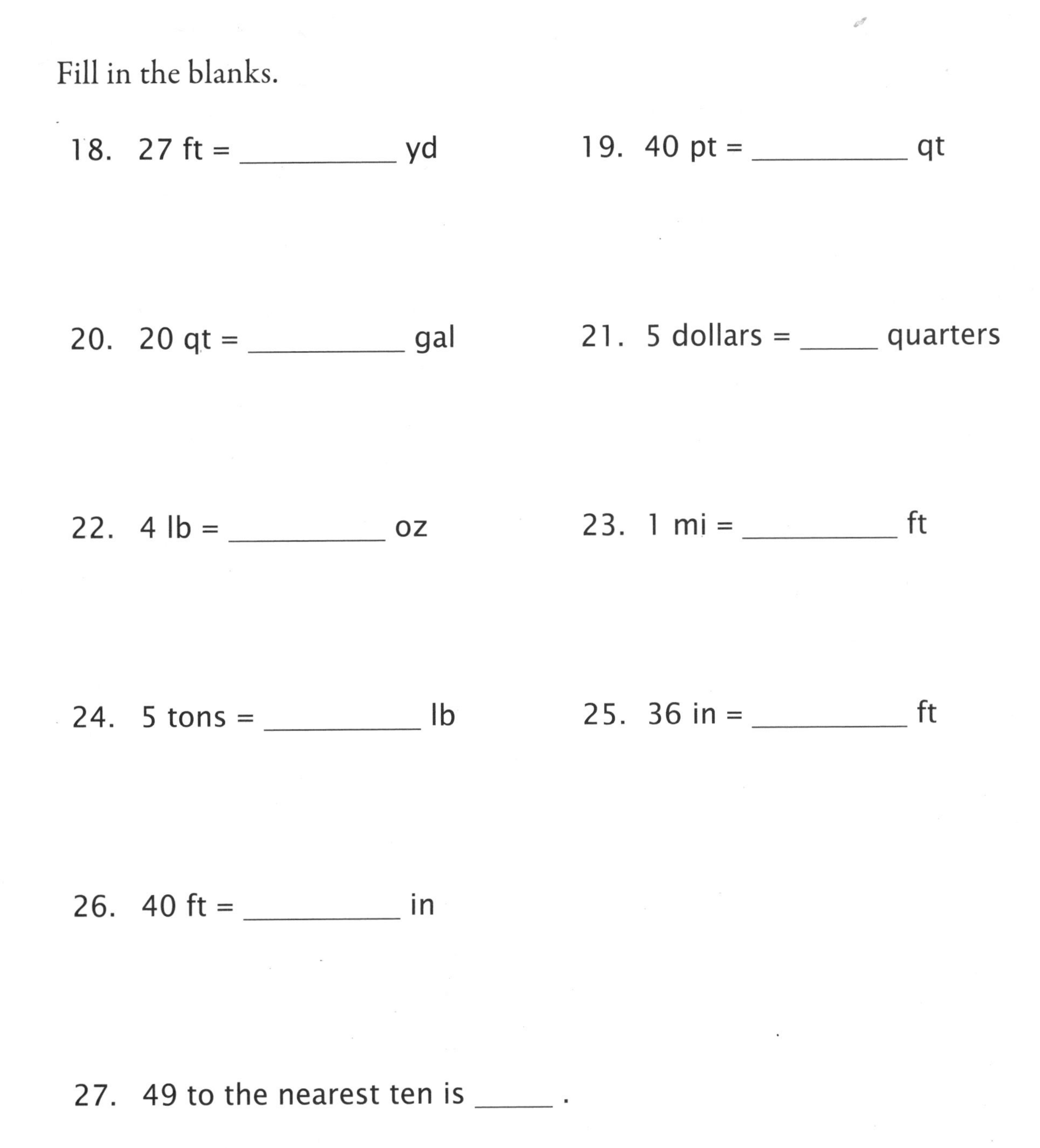

Fill in the blanks.

18. 27 ft = __________ yd

19. 40 pt = __________ qt

20. 20 qt = __________ gal

21. 5 dollars = _____ quarters

22. 4 lb = __________ oz

23. 1 mi = __________ ft

24. 5 tons = __________ lb

25. 36 in = __________ ft

26. 40 ft = __________ in

27. 49 to the nearest ten is _____ .

28. 4,009 to the nearest thousand is _____ .

29. 459 to the nearest hundred is _____ .

Solve.

30. $\frac{1}{3}$ of 12 = _____

31. $\frac{3}{7}$ of 21 = _____

32. $\frac{5}{8}$ of 32 = _____

Find the denominators and numerators of the rectangles.

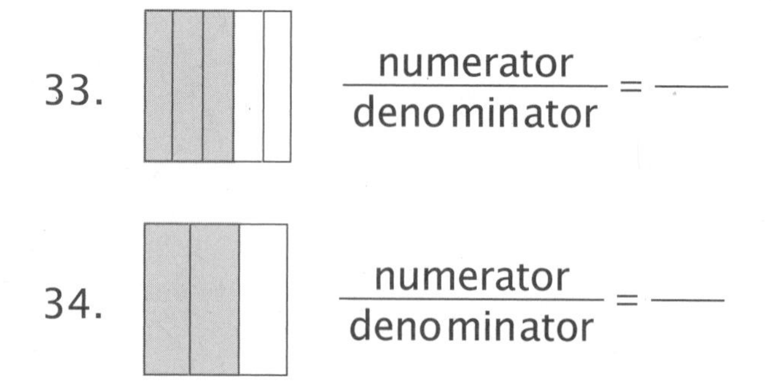

33. $\frac{\text{numerator}}{\text{denominator}}$ = ___

34. $\frac{\text{numerator}}{\text{denominator}}$ = ___

35. Write in standard notation:

2 x 1,000,000,000 + 5 x 100,000,000 + 4 x 10,000,000 + 3 x 1,000,000 + 9 x 100,000

______________________________________________

36. Find the average of the numbers: 5, 12, 13, 21, 24 _____

37. What number is represented by the Roman numeral MMCLVIII? _____

38. Write the given date with Roman numerals: 1975 _____